W9-ABI-163

Alice in the Land of Plants

Yiannis Manetas

Alice in the Land of Plants

Biology of Plants and Their Importance for Planet Earth

Yiannis Manetas
Department of Biology
University of Patras
Patras
Greece

Translated by Clio Weber, Thessaloniki, Greece

Translation from the Greek language edition:
ΤΙ ΘΑ ΕΒΛΕΠΕ Η ΑΛΙΚΗ ΣΤΗ ΧΩΡΑ ΤΩΝ ΦΥΤΩΝ

Cartoons by Myrsini Maneta, Athens, Greece

ISBN 978-3-642-28337-6 ISBN 978-3-642-28338-3 (eBook)
DOI 10.1007/978-3-642-28338-3
Springer Heidelberg New York Dordrecht London

Library of Congress Control Number: 2012940412

Cover design: deblik, Berlin, based on the cover of the Greek edition

Printed on acid-free paper

Springer is part of Springer Science+Business Media (www.springer.com)

Dedicated to my first teacher, Niky Parousi, and to:

Constantinos Mitrakos
Hemming Virgin
George Akoyounoglou
Nikos Gavalas
. . .who pointed me towards the scientific route

Φύσις κρύπτεσθαι φιλεί
(Nature loves to hide itself)

ΗΡΑΚΛΕΙΤΟΣ

Everything should be made as simple as possible, but not simpler.

ALBERT EINSTEIN

Preface

Why did I write this book, and why should you read it?

I could state that my motive is that I know plants; I have been teaching their biology at the university for 30 years. However, no truth is absolute, and the preceding statement sounds unacceptably vain and boastful. So to restate: I know, as does the rest of the scientific community involved in the subject, just a small part—a nugget, I would say—of the biology of plants, the part that present-day methods allow us to see and the current theories allow us to assume. This book contains what we accept about plant biology at the dawn of the 21st century. Future generations will learn more.

Let us assume that a university professor's aim is not only to announce the results of his or her research to the scientific community and to distill knowledge of his or her field for students, but also to transfer this knowledge to the general public, which is always an interesting intellectual exercise. The hardest questions asked of a scientist are those that seem naïve on the surface, such as those asked by a child or by a curious, intelligent, and perceptive adult who has nothing to do with the subject at hand—or by an intelligent extraterrestrial who has studied life forms completely different from those familiar to residents of planet Earth. So, how would I explain the biology of plants to my children? Herein lies the intellectual challenge. Especially noteworthy is Einstein's maxim that a scientist has no right to claim he or she knows a subject in depth if he or she cannot explain it to his or her grandmother. Thus, the point of the exercise is to forget yesterday's experiment, to focus one's attention on the essence of things and not on the details, and to lay the foundation for understanding plants, which appear and behave very differently from animals and have shaped life on our blue-green planet. The goal of this book is also to share 30 years of plant study with readers so they can look at plants in a different—and friendly and entertaining—way.

Is the project realistic? Can science be popularised? Can the average person comprehend scientific achievements? One would assume the success of such a project depends, at least initially, on the author, who must directly address the appropriate audience. However, the audience must also be open, willing to learn, and ready to put its cognitive powers into action and its imagination at the service of learning. I believe science is simple; comprehending and generating new knowledge (i.e., scientific research) both require attributes humans already possess. Otherwise, scientific progress as a product of human civilisation would not be possible. What is required first and foremost is common sense; even more necessary, however, is the ability to overturn common sense and the capacity and readiness to discard accepted views and to refute theories when evidence no longer supports them. Moreover, what is required is curiosity about the world surrounding us, which we must comprehend to make the right decisions and to recognise our real place within it. Unfortunately, most humans abandon this innate curiosity before the end of childhood when the first bored, tired, and intellectually apathetic adult squashes us with the phrase "enough of the questions."

As a university student, I used many methods to silently grade my professors. One of these methods had to do with the way they answered questions considered naïve. Some showed deep knowledge of the subject matter. Others failed, panicked, and surprised me with answers clearly irrelevant to the question, a sign that they themselves had never wondered about it. Some honestly admitted temporary ignorance and were better prepared at the next lesson. Thankfully, only a few answered, "That, sir, you should already know," with enough sternness and annoyance to prevent similar questions from being asked in the future, lowering student confidence and increasing the distance between teacher and learner. No one remembers the names of these professors, at least not in a positive light.

Thus, all questions are legitimate, and the most fertile ones are those considered naïve. One of the writer's tasks is to raise such questions and attempt to give answers compatible with our current views on the functions, behavior, and roles of plants, but beware! Not all questions can be answered. Usually, we raise only questions that may plausibly be answered. The others are left pending, swept under the carpet, ignored until they emerge mercilessly when the accumulation of new knowledge returns them to the fore. Therefore, do not shoot the messenger; the writer does whatever is possible to provide you with answers, when answers can be given today, when a question can be formulated. Searching for the truth is no cakewalk. Theories are confirmed until they are refuted. Navigation was carried out successfully for hundreds of years, even though it was based on the view that the earth stood still. What we believe to be true and self-evident today might be shaken tomorrow, because "nature loves to hide" and does not reveal its secrets easily. Our way to approach the truth, which always seems so close yet always evades us, is to study and maintain an attitude of curiosity about and admiration for the world: "Wonder and doubt," as the Delphic exhortation goes. This is when science changes from an intellectual bogeyman to an intellectual game.

Although the reasons presented here may easily explain the author's possible motives, why should readers be interested in plants? This is one of the first questions any reasonable publisher would ask; it might well concern you before you put your hand in your pocket. Why a book about plants? There are several reasons for such a book. First, plants are organisms so different from us we need additional guidance to comprehend their idiosyncrasies. Second, plants have the only broad-scale mechanism for utilisation of a practically inexhaustible source of extraterrestrial energy. Third, plants comprise 99% of the living mass on the planet. Fourth, plants shaped, are shaping, and will continue to shape the earth's atmospheric composition and hydrologic cycle in a way that is compatible with and necessary for all living organisms. Without plants, life on land, at least in its present form, would be impossible. As the most resilient organisms, plants existed before us and will continue to exist after us. Our energy and metabolic state of affairs depends on them, yet public opinion and many professional biologists—unfairly—consider plants inferior and simplistic organisms with no visible behaviour or intelligence. Finally, there usually is more to plants than meets the eye.

Although this book originally was intended for the general public, *Alice in the Land of Plants* also may be of interest to students of biology or other related sciences. It also might be useful for biology teachers in secondary and primary education, as it might help fill gaps or provide exciting examples for teaching that may lead to students' early realization of the different lifestyle of plants and their importance to planet Earth. To this end, the extensive use of marginal notes of a questioning, maximal, or categorical nature throughout the book has multiple purposes. The notes highlight essential points, guide the reader through the text, stimulate thought and memory, and serve as a verdict or final judgment on the issue at hand. Together, they comprise a smaller book within the larger one that may be read separately.

To enhance the flow of the main text, reference citations have been avoided. However, extensive general and specific bibliographic references appear at the end of the book.

I thank Dr. K. Zeliou, Professor G. Psaras, and Assistant Professor Y. Petropoulou, my colleagues at the University of Patras, as well as K. Chassourou, my colleague in secondary education, for the time they sacrificed to read part of the book and offer useful comments; Professor C. Thanos (University of Athens), Dr. T. Kokkoroyiannis, and C. Weber for critical comments incorporated into the second Greek edition; and the University of Crete Press, especially its director, Professor S. Trachanas, Ms. D. Daskalou, Head Publishing Editor, and Mr. N. Koumbias, Scientific Editor—the first two for their positive disposition toward the book's publication and the third for his particularly apt comments. Finally, I thank the students throughout my 30 years of teaching who always—especially during periods of educational disappointment—made me call out aloud, "Worthy is the price paid."

Yiannis Manetas

Acknowledgments for the English Edition

The English edition of *Alice in the Land of Plants* would not have been possible without the additional help and encouragement of some friends and colleagues. My sincere thanks go to Dr. Theodoros Kokkoroyiannis (Athens) and Prof. Tasios Melis (Berkeley) for the first and the final push, respectively; Clio Weber and Colin McCullogh for the excellent translation; Vicar Des Williamson for reading part of the English text; my editor at Springer, Dr. Christina Eckey, for the fine blend of enthusiasm and professionalism; Dr. Jutta Lindenborn for the excellent project coordination; Georgette Forgione for editing the text; and, finally, to my wife and colleague, Dr. Yiola Petropoulou, for her patience, and my daughter, Myrsini, for giving form to *Alice*.

Contents

Chapter 1
Introduction

Plants are no Less Complex than Animals: They are Just Different

If the average person were asked how important plants are, it is more than likely he or she would not underestimate them. Who could dispute their importance as a source of food when 35 % of the dietary protein in the developed world and 80 % in the developing world comes directly from plants? Indeed, all our food comes from plants because they are the base of the food chain. Moreover, 25 % of all medicines in the West include at least one plant extract, and this percentage increases sharply if calculations include traditional medicine in the Third World. Those better informed might add that civilisation – whether in antiquity or in the industrial era – was based on plants. As for literary and scientific masterpieces, these were recorded on plant products, such as Egyptian papyrus and Chinese paper. Those who are more sensitive might refer to the aesthetic value of plants or the deep emotions aroused by their wide range of colours, hues, and scents. Others may remind us of the delight plant condiments add to food, the merriment following a glass of wine among friends, or the ecstatic harmony of a violin made from the right wood.

The significance of plants is not limited to their use by humans

One could carry on *ad infinitum* along these anthropocentric lines and write page upon page about the usefulness of plants; however, that is not the aim of this book. The aforementioned list of various plant uses refers to only a few hundred plant species, a tiny minority, that have been "tamed" and cultivated by humans or that provide their products and services as indigenous wild species. We share the planet with at least 260,000 other, insignificant plant species that need support and are truly significant merely because they exist. Even if people are oblivious to them, such species have shaped and continue to shape the state of affairs on Earth and

Plants keep shaping the biological state of affairs on planet Earth

will do so as long as the planet exists. This is because plants are the only organisms on Earth that can easily exploit a virtually inexhaustible, extraterrestrial energy source, and because the process selected for this successful exploitation, photosynthesis, has to a great extent determined the climate, the composition of the earth's atmosphere, the carbon cycle, the water cycle, food production, and so on. Photosynthesis shaped the history and evolution of life to an extent that we have only begun to grasp in recent decades. It is easy to imagine a world that is exclusively vegetal, without humans or even animal species, such as those that exist or used to exist on Earth. However, it is impossible to conceive the opposite. In their search for elementary life forms in the universe, the first thing scientists look for is water. Small pockets of bacteria, with limited exploitation of local chemical energy sources and an equally limited capacity to propagate far from these sources, may be invisible at first glance, buried in the depths of underground or surface water masses. This hardly looks like the blue-green planet. A rich terrestrial life as complex as that encountered on Earth, covering the whole available surface, is inconceivable without some form of plant life. During nine tenths of our planet's history, life was limited to water while the earth's terrestrial surface was a grotesque, inhospitable lunar-like landscape. Only after the gradual (yet exceptionally speedy in geological terms) colonisation of the earth by plants did life as we know it today become possible. Because of this colonisation, we have inherited a world with millions of animal species that find habitat, food, and protection thanks to the 260,000 plant species in every possible nook and cranny of the planet. This precious heritage deserves our appreciation and protection, but first we must become acquainted with it.

A world without animals is possible, but a world without plants is inconceivable...

...particularly on dry land

Let us, therefore, get to know the food and protection providers

Natural versus Human Selection

Wild and cultivated plants or natural and human selection

This book is not about cultivated plants, which in any case are quite different from their wild ancestors. This difference lies in the fact that wild plants are the result of natural selection whereas cultivated plants are the products of human selection. Of course, the latter originated from wild ancestors, but human intervention in their evolution has been so effective that today they bear little resemblance to their ancestors, which in many cases are unknown. Human selection is aimed mainly at changing the natural properties of a species to serve human purposes rather than to benefit the plant itself. Cereals are a typical example. Like all plants, the wild ancestors of wheat, barley, and maize scattered their grain (contained in their fruit, i.e., their ears) away from the mother plant, so the offspring would grow at some distance. In this way, vital territory increased and the new plants were not faced with competition from their progenitors. It would be a great reproductive failure if the grains did not leave the mother plant: What would be the point in their growing on it? Grain dispersion is controlled genetically by several genes that are activated when the seed has matured. In some cases, however, random mutations may result in unsuccessful dispersal, causing the seed to remain on the plant's ear, where it would be pointless to germinate. When humans recognised the nutritious value of cereals, they started collecting their grains and, obviously, collected those still on the plant, which was much better than picking at the soil like hens. This is how the first grains were selected. They were mutants, probably the first mutants consumed in history. Nonmutated grain had already fallen to the ground, forming the basis for the next harvest.

How were cereals selected?

When humans were still food gatherers, there was no point in keeping leftover grain, that is, grain that had not been consumed. Therefore, after they ate their fill, they got rid of the rest because refrigeration did not yet exist to preserve food. Eventually, some of the more observant people among this group of early humans noticed that these same nutritious plants were growing at sites where human rubbish had been disposed.

The First Gardens

Contemporary common sense dictates that if one wants to confirm that a phenomenon is not an accident or a miracle, he or she must repeat the manipulation to determine whether the phenomenon occurs again. In other words, one must perform a controlled experiment. Although this was not an easy task for our primitive ancestors, some charismatic individuals did perform an experiment, resulting in the first gardens. Now, instead of foraging for wheat, they could cultivate it on site, transforming themselves from food gatherers to cultivators with a fixed abode. What did our ancestors cultivate? Obviously, mutant individuals. Unknowingly, they selected plants that could not reproduce on their own, as the grains remained fixed on the mother plant. As a result, wheat has not been able to survive without humans since then; it has become symbiotic and cannot exist independently. It separated from its wild ancestor; it can grow as a plant but cannot leave offspring behind unless humans sow its grain.

The first gardens of humanity

Most cultivated plants are, in a sense, mutants

Other cultivated plants had similar destinies. The wild ancestors of lettuce, for example, have hard, bitter leaves, which serve as a defence system to protect the species from being overconsumed by its predators. This property is very useful for the plant; it was tested and naturally selected to achieve a balance in which predators can avoid famine while the plant species avoids extinction. Occasionally, random mutations occur in the genes responsible for tough cellular walls or for the biosynthesis of bitter phenols. Such mutations may lead to the appearance of plants with limited mechanical and chemical defences. These mutants cannot compete with their bitter, tough relatives unless they are selected by humans. Of course, such disadvantaged and handicapped lettuces are preferred not only by humans, but also by other herbivores and pathogenic microorganisms, which humans must eradicate with fungicides, insecticides, and other cultivation practices. If left without protection, such lettuce types may not reach reproductive age, nor will they leave many offspring behind.

Human selection produced plants incapable of surviving on their own. Since then, they have been symbiotic with humans...

For the aforementioned reasons, cultivated plants do not escape easily from their crops and are unlikely to be encountered growing on their own. They are strictly symbiotic with humans. If people decided to feed on pills, such plant species would disappear within a few years and their present-day artificial habitat would soon be overtaken by wild species. Cultivated plants are like pet dogs, aquarium fish, and canaries in cages: to survive, they need an artificial ecosystem provided by humans; they would not survive natural competition. They are biotechnology products with an expiration date, a topsy-turvy selection. Their significance is of concern only to humans, in a direct and exclusive way. They will collapse as soon as humanity collapses. They are useful for the here and now but of no importance for life's continuum.

...and cannot easily escape from cultivated crops

This book, therefore, does not examine the plants humans spend time cultivating, consider valuable, and describe as useful. Rather, it focuses on the insignificant plants, those we pass by every day without noticing, those we look at without seeing, those we step on with no protest on their part, those we cut down to improve the view from our windows, those we consider unworthy of being viewed (although we *should* consider them worthy), those that existed before us and will continue to exist when we are gone. If these plants all have a role

This book studies wild plants and their functions, which are important at the planetary level

to play, both as a group and as individuals, if they have a significant function that escapes us, if the state of our planet would be completely different if these plants did not exist – which is what this book posits – then it might be worth paying greater attention to them. The prerequisite for appreciating these plants, of course, is that we become acquainted with them.

Plants versus Animals

When contemplating Alice's entrance into Wonderland, certain intellectual difficulties arise that also must be faced when encountering the world of plants. Like humans, plants are organisms, yet they are quite different from us. Plants also differ from organisms that, because of their similarity, are more familiar to us (i.e., animals). It is understandable that humans easily comprehend animals, with which we share a basically similar lifestyle. We are familiar with this lifestyle because each person's body – including its form, functions, and behaviour – is an inevitable part of his or her experience and is needed to maintain health and physical conditioning. However, the structure, functions, and behaviour of plants are fundamentally different; therefore, comprehending them requires a mental shift. The most effective way to think about plants is to compare animals (which are more familiar to us) with plants (which are not) to identify basic similarities and, more importantly, the basic differences that impose a different structure, organization, development, function, and behaviour.

To enter the world of plants, we need to get rid of our anthropocentric view of things

If the average reader finds the aforementioned difficulties justified, or even self-evident, he or she would expect them to have been overcome by experts, such as students of biology and related disciplines. Unfortunately, this does not seem to be the case.

The relevant literature appears to consider the basic organizational and functional philosophy of the plant world to be self-evident, although it is not. This philosophy is bypassed to leave room for otherwise useful details that, however, are not examined within the context of the essential pattern of plant behaviour. In other words, the general prevailing sense is that because plants are not structured, are not organized, and do not behave like humans, they are incomplete organisms and so are unworthy of our curiosity, observation, and study, as opposed to animals or, more importantly, to humans. In this view, 99 % of the earth's biomass is of no consequence and hardly deserves scientific attention.

Plant Blindness

Plant blindness: a mental attitude that condemns plants to discredit

A decade ago, some biologists in the United States tried to scientifically explain the low esteem and modest interest humans have for plants. They coined the term *plant blindness* to describe an attitude that fails not only to explain the life of plants, but also to even recognize their existence around us. Since then, a series of Gallup polls of students, biology teachers, and the general public have shown that the reputation of plants is disproportionately low in relation to their significance as living organisms. Yet scientists are still at odds as to the reasons behind such blindness. Some report there is a physiologic explanation for plant blindness: the human eye receives 10 million information bits per second, but the brain processes and focuses attention on only a few (around 10). This unconscious choice on the part of the brain is based on its detection of three characteristics of objects: movement (plants do not move), bright colours (although flowers attract attention, they appear only seasonally), and potential danger (who is afraid of plants?). Because we cannot overcome our inherited

Physiologic basis for plant blindness

characteristics easily, this physiologic view provides us with an alibi. In other words, plants are objects we see but do not pay attention to; they do not arouse our curiosity.

Others, however, maintain that the physiologic basis for such blindness is reinforced by social and educational prejudice. A biology teacher who is asked to describe a basic biologic function using an example in most – if not all – cases will choose an animal or, better still, an anthropocentric example, probably because he or she is more knowledgeable about it. Perhaps animals are more familiar to the biology teacher because during the teacher's studies, he or she did not make the necessary mental shift to see plants in a different light.

Social and educational bias

In any case, this blind attitude towards plants, whether inherited or acquired, clouds one's judgment and degrades plants to organisms that deserve less attention. Yet, the greatest fathers of modern biology were inspired by plants and dedicated the main part of their scientific work to them. One of the great naturalists of all time, Alexander von Humboldt (1769–1859), was (among other things) an enthusiastic collector of plants and a keen observer of the effect of climate on the geographic distribution of plants. The father of the theory of evolution, Charles Darwin (1809–1882), supported his views by observing both plants and animals, but his descendants isolated and highlighted the example of finches on the Galapagos Islands. Few are aware that the greatest part of Darwin's scientific work concerned plants. Gregor Mendel (1822–1884), the father of classic genetics, formulated the laws of heredity based on his experiments with peas. What happened in the meantime to lead us to the current state of affairs? Why are universities around the world offering fewer and fewer courses on plants? Why is funding for plant research decreasing? Why do biology books at secondary schools and universities contain fewer chapters dedicated to plants as each new edition

The most important fathers of modern biology studied plants carefully

Why do the descendants of scientists ignore plant studies?

is published? Why do students rarely choose to study plants during their postgraduate studies? Why does the average person consider plants "third-class citizens?"

In the struggle for survival among scientific branches the winner is: molecular biology

It is not the aim of this book to analyse this phenomenon, and the author might not be fully qualified to do so. However, plant biologists may not have marketed their products effectively, and they are not the only ones. Many other (equally interesting) objects of biological study are gradually sinking into obscurity, failing to counterbalance the prevalence of molecular genetics, especially its applied branch of biotechnology. The latter indisputably has exploited its fundamental success at the basic research level to promise, probably prematurely, radical solutions to human health and nutritional problems. There is a widespread impression that the initial and justifiably sensational optimism of this discipline soon led to the childhood disease that almost inevitably accompanies every quick success story: a sense of supremacy and a loss of perspective.

Popular Science

Comments on the communication of research

In the Middle Ages, alchemists convinced rulers – because they were convinced themselves – to provide them with the funds they needed in their quest for the philosopher's stone, which would turn humans into immortal beings and base metals into gold. This activity proved to be a futile attempt that consumed both wealth and human effort. However, the alchemists' pursuit had important results. During the initial, heroic period, a wealth of knowledge was generated about the nature of metals, along with methods for studying them. Although the philosopher's stone was never found, the research methods discovered were useful in the ensuing development of chemistry. In time, the number of crooks and charlatans grew so much that the rulers were forced to ban the practice of alchemy.

Although things might not be exactly the same today, *mutatis mutandis,* several similarities exist. Research no longer is funded by arbitrary rulers accountable only to themselves but by governments whose leaders want to be reelected or by private enterprises accountable to their shareholders. Still, supporters of science today, as in the past, expect some sort of compensation – reelection or direct financial gain; therefore, they must be convinced their investment will pay off. Medieval alchemists had to convince only the local rulers, at the risk of losing their heads if they failed. Although modern scientists do not face such a drastic fate if their research ends in failure, they have to convince more people, that is, they must sway public opinion in their favour. Today, public opinion is influenced by the media, and when one secures a few minutes of publicity, he or she achieves the glamour of being famous and is automatically labelled an "expert" by the public. Countless times a day, we hear the statement "I saw it on TV." We do not hear "I read it in a book," or "I know from a reliable source," or "I researched it on my own," but simply – and effortlessly – "I saw it on TV" (so it must be true!). A scientist appearing in the popular media may fume when his or her statements are distorted by journalists, yet the exposure may lead to a publishing contract for his or her book, the ability to influence public opinion, and recognition of his or her views.

The glamour of being a "celebrity"

Experts in molecular biology and biotechnology played this communications game well: they built their myth; marketed their product effectively; promised signs and miracles; monopolised the interest of the public, students, and scientists; and marginalised the other branches of biology. As far as public opinion is concerned, biology is now synonymous with DNA, which has become a household term. Any characteristic considered inherent is now said to be "written in one's DNA," and there are people preparing to tell us our

Myths and promises...

destiny and predetermine our future based on a genetic sequence. Although such a sequence might be known mechanistically, more serious and discerning molecular biologists contend that what we do not know – the how and why of genomic regulation and function – is infinitely more than the smidgen we do know.

...when the degree of ignorance is many times that of knowledge

This book's aim, however, has nothing to do with research communication; it has to do with the fact that plants are considered inferior organisms. To the physiologic or sociologic explanation of this attitude towards plants, I would add the view – justifiable at first glance – that plants are simpler organisms than animals.

Complexity

Criteria for the complexity of organisms

What determines an organism's complexity? A system is complex if the whole cannot be explained based on the properties of its components. Although these properties are easier to analyse, their sum is hardly ever sufficient to explain the whole, because the parts are interdependent in various, and not always predictable or easily accessible, ways. Therefore, one of the parameters of complexity relates to the organism's size and form. Obviously, the more cells, tissues, and organs an organism has, the more complex it is. In this sense, microorganisms must be simpler than other organisms. Within each distinct kingdom, complexity should increase with size, that is, with the number of cells and their need to communicate with one another. If one adopts this view, a horse is more complex than a worm and a plane tree more complex than a dandelion plant. However, is this actually true?

Size and complexity

Behaviour and complexity

If, for a moment, we ignore form and examine behaviour – that is, an organism's action and reaction to environmental challenges – we reach a similar conclusion. Animal behaviour is more complex than that of plants, which, after all, do not move – but is this actually the case?

Although it may be premature at this point for the reader to believe plants are no less complex than animals, the hope is he or she will be convinced through further reading. However, it must be stated at the outset that plants have all the component biological properties that characterise behaviour. They perceive the environment, they literally measure it, they record and respond to stimuli, they regulate the chemistry within their bodies and the physics and chemistry of their surroundings, they feed in a characteristic and self-sufficient manner, they perform cellular metabolism at their discretion (or, to avoid such teleologic views, in accordance with their developmental programming, as this is modified by the environment), they develop, they differentiate, and they reproduce. As for their apparent lack of motion, indeed they do not get up and leave when threatened – for example, if they become too hot or cold – they simply change their properties where they stand to cope with the threat; this is typical plant behaviour. Although this behaviour pattern would be impossible for humans or animals, it is successful for plants, as proven by their presence everywhere on the planet. Yet, plants do not completely lack movement. Plants do move in a certain way and often enough to fully serve their needs, but we cannot see their motion because we are not used to perceiving such slow movement. One may detect plant motion by careful observation and technical means, as well as through an effort to ignore the anthropocentric concept of motion, which says, "Flee when threatened, approach if you expect a reward." The penetration of roots towards selected soil regions is spatial movement, as is the upward growth of the shoot towards light – and there are more examples, as the following pages reveal.

Do plants have "behaviour patterns"?

And yet, in their way, they move

Even if we accept that plants are indeed complex organisms, how do we respond to the question of whether they are more or less complex than animals? Is there a quantitative criterion for complexity? For this, we can turn to molecular genetics and its central

Molecular criteria of complexity

doctrine that the form and functions of an organism are determined and controlled by material informational bits called genes. Recent technologic advances have made it possible to determine the number of genes for some organisms, including humans. It would be reasonable to expect that more genes exist in more complex organisms. For decades this reasoning was accepted although it could not be proven experimentally that the number of genes, generally speaking, is lower in plants, higher in invertebrates, and still higher in vertebrates. However, modern molecular biology has confirmed that this is not necessarily the case. Although determining the number of genes is a difficult and controversial task (as indicated by frequent revisions), it is generally accepted that around 21,000 to 23,000 genes exist in humans. *Anopheles gambiae*, the mosquito that carries malaria, is indisputably a less complex organism, with around 14,000 genes. Is this difference in the number of genes a measure reflecting the difference in complexity between a human and a mosquito? Are the 7,000 additional genes enough to account for the human condition? Is a human only four times as complex as the microbial pathogen *Escherichia coli*, with its 5,500 genes? Molecular biology keeps surprising us: a dog has as many genes as a human, whereas a sea urchin has a few more (24,000). The humble roundworm *Caenorhabditis elegans*, which is only 1 mm long, has as many genes as the so-called capstone of creation. What about plants? *Arabidopsis thaliana* has 28,000 genes. Is this a giant of the animal kingdom? No, it is an herb that is scarcely 20 cm in height. Rice has about twice the number of genes as humans and the poplar tree even slightly more (44,000 genes). Therefore, we need to revise our views of what exactly biological complexity means or accept that complexity is not related (or not related only) to the number of genes.

How is it possible that a sea urchin has as many genes as a human being...

...and rice has twice as many genes as humans?

Of course, molecular biology has provided more important information, for example, that the protein products of some genes function as switches that awaken certain genes or lead others to dormancy. Technically, these products are called *transcription factors*. Might this be where the quantitative expression of complexity lies? It is reasonable to assume that organisms that appear more complex have a greater need for gene function regulation. Yet, this does not seem to be where the secret lies. Humans have around 2,000 genes that codify transcription factors, whereas the humble *A. thaliana* plant has around 1,500. Nevertheless, it is known that protein products of genes are edited later so that a gene ultimately produces many more than just one protein. Therefore, the key might well lie in the maximum number of proteins produced (i.e., the proteome). Nevertheless, the comparison favours plants.

Transcription factors and post-transcription modification of proteins: plants prevail on all counts

One may wish to leave the issue open, expecting the question as to which organisms are more complex to be answered in the future. However, there is at least one reasonable explanation for the apparent wealth of genes, transcription factors, and proteins in plants. This explanation may be related to plants' immobility and their need to face head on, with no means of escape, seasonally changing environmental conditions, attacks by predators and enemies, and competition from neighbouring plants. These issues are discussed again later. For now, following the fluidity of criteria imposed by recent discoveries in molecular biology, the least we can say is that plants are not necessarily less complex than animals. Moreover, plants are not less successful as organisms; therefore, they are as complex as their lifestyle and life pace dictate. They simply are different. The aforementioned phrase "they are not less successful" actually may be an understatement. In effect, plants are *more* successful – on an evolutionary and geologic scale. The average time plant species have been present on the planet is many times that of animal species.

Is it immobility that requires more genes?

In other words, plants are no less complex; they are just different and, in evolutionary terms, more successful

Animal species come and go and have a shorter "sell-by" date. If we could travel to the distant past or distant future, we would meet more or less the same plant forms, but quite different animals. The reasons should become clear in the course of this book.

Chapter 2
Basic Plant Organisation: How it Differs from that of Animals

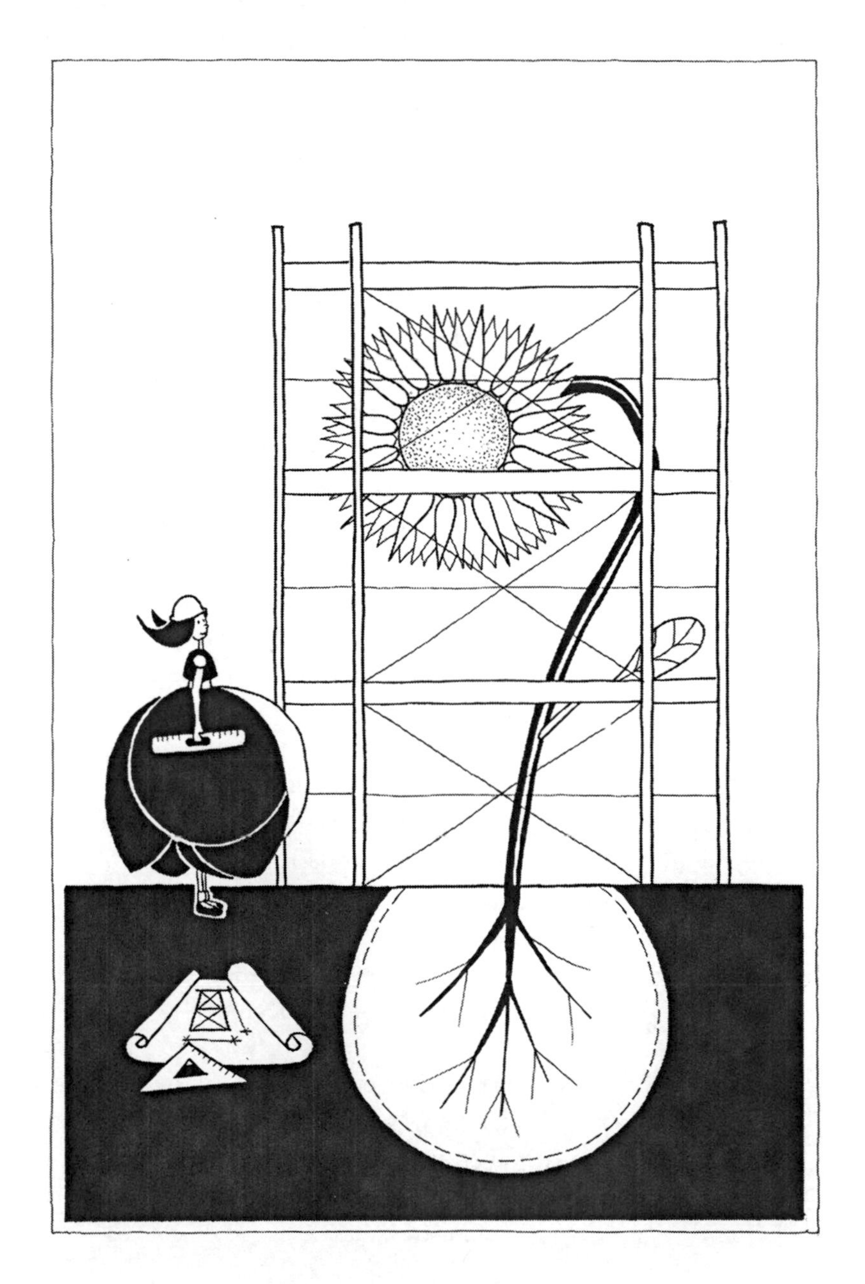

Plants are not simpler organisms – they are just different

First, let us accept that plants are no less complex than animals – they are just different. Therefore, to understand the functions and behaviour of plants, we need to be aware of their basic similarities and, mainly, their basic differences compared with animals, which are more familiar to us. At the cellular level, the similarities prevail. Plants are composed of cells containing a nucleus that stores genetic information; mitochondria responsible for cellular respiration; an endoplasmic reticulum of membranes to help complete three-dimensional protein structures and their endoplasmic transport; Golgi apparatuses to classify, "pack," and dispatch where they should go within the cell (or excrete from the cell) groups of biological macromolecules; and an internal cytoskeleton to maintain the cell's form and to rearrange the relative positions of internal organelles. All these organelles are common in eukaryotic organisms, that is, all organisms except bacteria. The latter do not contain organelles in their cells; however, that does not mean they are imperfect organisms. Plant cells also perform protein synthesis, cellular respiration, and intermediary metabolism, in almost the same primordial manner as all other organisms. Furthermore, however, plant cells have chloroplasts that are responsible for photosynthesis. They also have a cell wall, a hard external skeleton that keeps the shape and volume of their cells, tissues, and organs more or less stable. Fungi and bacteria also have cell walls. Finally, plant cells – as opposed to animal cells – contain the so-called vacuole, a relatively large reservoir that occupies 70 % to 98 % of the interior cellular volume and, in effect, functions as a storage space. Except for chloroplasts and the photosynthetic function, which are missing from animal cells, all other structural and functional differences between plant and animal cells are minor compared with their similarities.

Most cell organelles are common in animal and plant cells…

…however, plant cells also include chloroplasts, cell walls and a huge storage area called a vacuole

However, the philosophy of the organizational structure of plants at the level of the whole organism, as well as the behaviour, is totally different. Let us examine these differences by asking the right questions.

Plants are Nutritionally Self-sufficient

What do plants eat? This question is somewhat misleading. Words have been invented by people to describe the world as they perceive it, and quite often they cannot apply these words equally to other systems. If one is asked what humans eat, the answer might be "whatever they put in their mouths and swallow." We know, of course, that food (whether of plant or animal origin) is processed in the digestive tract, where useful ingredients are separated from useless ones. The former are absorbed and the latter excreted. So, in effect what humans are seeking from food are the proteins, lipids, sugars, vitamins, and mineral nutrients it contains. The first two groups of substances, proteins and lipids, are broken down further into smaller components (amino acids and lipids of lower molecular weight) to facilitate absorption.

Plants and animals need exactly the same organic substances...

The ingredients absorbed are used in three ways: first, as building blocks to construct the biological macromolecules typical of our species. For example, proteins are made up of amino acids. Although in theory there are an infinite number of amino acids, only 21 of them participate in the protein synthesis of all organisms, regardless of their evolutionary level, from bacteria to plants to humans. Milk or cereal proteins, however, are not the same as human proteins: they differ in the proportion of amino acids they are made up of and the sequence in which they form bonds with one another in the protein. The ratio, sequence, and total number of amino acids give every protein its structure, volume, and properties, which determine two main things: its specific action (e.g., as an enzyme) and the position it occupies in the cell. Both these characteristics are important, the former self-evidently and the latter because cell ingredients cannot be scattered in a disorderly manner; they must be located where they can perform their function optimally and be regulated properly. With regard to digestion, the organism first breaks down food protein into its component amino acids, which it then rearranges in a specific way to synthesise exactly what the organism needs. This process is like a child taking Lego bricks apart from a random "house" to build one he likes better. Animals always use ready-made "bricks" – amino acids – whereas plants construct their own amino acids from simpler and cheaper materials, as is shown later.

...which they use to build their bodies, repair damage, and retrieve the energy contained in them

The second way of using food is as an energy source. All organisms need energy to develop and maintain themselves, and those that move need quite a bit more. Energy is necessary to build the new house with the bricks available, as well as to repair any damage that inevitably occurs. The organic substances humans take from food, particularly fats and sugars, are rich in energy, because a great deal of energy was used in their synthesis; this energy is contained in the substances as long as their structure is preserved. A molecule of glucose, for example, contains tens of times more energy than the two simpler molecules from which it can be synthesized: carbon dioxide (CO_2) and water (H_2O). Of course, if we looked for the source of the wealth of energy contained in glucose (as well as any other organic molecule included in our food), we would not find it any dark places. All this energy comes from the sun, and plants play the role of mediator. The transformation of solar energy into chemical energy that is useful for living organisms is called *photosynthesis*.

The energy contained in almost all organic substances on the planet is of extraterrestrial origin. Plants are mediators in transforming solar energy into useable chemical energy through photosynthesis

The energy contained in organic matter is retrieved by breaking down this matter. A common way is to temporarily increase the temperature of organic matter to the point at which it reacts spontaneously with atmospheric oxygen (O_2). What follows is known as fire. Oxygen oxidizes organic substances violently and without control, returning them to CO_2 and H_2O while their energy is released as heat. Basically, this is what organisms do as well: they oxidize the organic substances contained in their food to obtain energy from within. However, this oxidation is not performed "violently and without control," and the organism is not set alight because oxidation is gradual and regulated, meaning it allows the organism to successfully obtain part of the energy contained (in the case of glucose, up to 56 %) and transform it into useful chemical energy to support growth, maintenance, and – if it moves – motion. The remaining energy inevitably is transformed into heat, as no energy transformation is 100 % efficient. However, because the lost energy is released gradually, not violently, the organism is heated without burning. The biochemical process of obtaining the energy contained in useful organic food substances is called respiration. It takes place in all aerobic organisms (i.e., those that not only are unharmed by O_2 but benefit from it) in almost the same biochemical manner, with the substantial participation of organelles called *mitochondria*. Respiration occurs in all body cells of all aerobic organisms, and the oxidizing agent is atmospheric O_2. Therefore, if one wonders why we cannot live without oxygen, the answer is that we need it to oxidise part of our food gradually and in a controlled manner to obtain the necessary energy and to transform it into useful chemical energy. The oxygen is obtained from the atmosphere through inhalation (breathing) and travels to the lungs and then, via haemoglobin, to all body cells. In plants, the epidermis is not a significant obstacle to O_2, which is taken up by almost all parts of

Retrieving the energy contained in organic substances takes place through their gradual and controlled oxidation, in a series of biochemical reactions called respiration

Respiration at the cellular level is the same in all aerobic organisms

Why do we need oxygen?

the plant body, particularly through tiny orifices on leaf surfaces known as *stomata*.

The third way of using food was already mentioned earlier and by now should be self-evident. Food is used for repairing damage, which requires not only energy but also new materials to replace the damaged ones.

The fundamental difference between animals and plants is that animals need ready-made, condensed, and packaged food, whereas plants manufacture their own food using simple, cheap, and naturally abundant inorganic substances. Plants are nutritionally self-sufficient.

With the knowledge that plants and animals need exactly the same materials – that is, their food consists of proteins, sugars, lipids, vitamins, a handful of minerals, and water – the initial question can be reformulated. If these items are defined as food, then there is no substantial difference between animals and plants, because they both need the same things. The fundamental difference, however, lies in the fact that plants do not eat but manufacture their organic food using the simplest materials that exist in nature. On the contrary, animals obtain ready-made food by eating plants or other animals that feed on plants. Plants are food manufacturers, whereas animals are food consumers. Because plants manufacture their food from simple raw materials, they are called *autotrophic* organisms, whereas animals are *heterotrophic*. From a nutritional point of view, plants are self-sufficient, whereas animals are dependent on plants.

What are the raw materials for manufacturing plant food?

What are these raw materials plants use to manufacture food? In other words, what is – *mutatis mutandis* – the equivalent of what humans swallow? What do plants put into their systems to manufacture their food? Water and the necessary minerals are common to both autotrophic and heterotrophic organisms. Plants obtain water and the minerals they need from the soil through their roots. The basis of biological molecules, though, is carbon, and its capacity to form bonds with four other chemical elements (whether the same or different) concurrently allows it to be part of a theoretically infinite number of organic substances. Whereas animals use as their carbon source some complex biologic material, which they need to consume ready-made, plants use as their carbon source the omnipresent atmospheric carbon dioxide (CO_2) and transform it into organic substances. However, this process requires energy. It is the energy stored in organic substances that subsequently is used by heterotrophic organisms that retrieve it. The energy source used by plants is of extraterrestrial origin and is sustainable. Through photosynthesis, plants bind solar energy, stabilise it in the form of chemical energy, and use it to synthesise organic substances. Plants share this capacity with some photosynthesising bacteria and algae. The reason algae are no longer considered plants – although the two groups of organisms have several similarities – is discussed elsewhere in this book.

Photosynthetic cells are factories that take up simple, unprocessed materials (CO_2 from the atmosphere, water and minerals from the soil) and use energy from the sun. They produce processed, good-quality food that is exploited by the entire biosphere.

Therefore, plants may be thought of as factories that take in simple, unprocessed raw materials (water, CO_2, minerals) and solar energy, and produce ready-made, processed, and good-quality organic food for the rest of the biosphere. Moreover, like all factories, in processing raw materials the photosynthetic factory produces by-products, that is, waste that must be removed. The main such product is oxygen, which is as useful as food. As discussed later, the O_2 produced by the first photosynthetic organisms appearing on Earth was an undesirable pollutant, even for its own manufacturers. Thus, hundreds of organisms that were anaerobic could not tolerate the presence of this pollutant and must have disappeared. However, in a splendid evolutionary move – after numerous attempts to evade, neutralize, or tolerate the pollutant – a means was found to make the pollutant useful through the process of aerobic respiration. In other words, a way was found not only to exploit an inexhaustible energy source such as the sun and to transform an exceptionally cheap and abundant source of atmospheric carbon into organic matter, but also to successfully use the polluting by-product. The result was perfect sustainability. For every six

Oxygen is a by-product of the process, and it is released into the environment

Oxygen, as a photosynthetic by-product, initially was a dangerous pollutant that destroyed the prevailing anaerobic organisms

Consequently, nature found a way to render this pollutant useful: it has been and is still used as an oxidant during aerobic respiration

molecules of CO_2 photosynthetically transformed into glucose, six molecules of O_2 are produced. For every molecule of glucose consumed during respiration, six molecules of O_2 are needed and six molecules of CO_2 are produced. The chemical process of respiration is exactly the reverse of photosynthesis.

Why is the chemical composition of the atmosphere more or less stable? How do plants warrant this essential stability?

When the global photosynthesis of plants, algae, and cyanobacteria is equal to the global respiration of all organisms, then the atmospheric levels of CO_2 and O_2 remain stable, at the levels known and desirable to maintain our civilisation. If this balance is disturbed, if one process does not counterbalance the other, then climatic changes occur. Later, this book discusses the fact that ever since they appeared on the earth, plants have served as safety valves to maintain the composition of the atmosphere and the climate in a manner we have only recently, in the last decade, started to appreciate. The question is whether this safety valve will be effective enough to avert the climate change imposed by recent – hardly sustainable – human activities.

Nutritional Self-sufficiency Renders Movement Redundant and Determines Plant Design and Construction

Nutritional self-sufficiency allows plants to be stationary

Regarding the basic differences between plants and animals, some of these stem directly from the basic difference in the manner of nutrition. Animals need to move to find their food; they must hunt, graze, perhaps look for carrion, or go to the supermarket. When food becomes scarce in one location, animals move to another. Plants' "food" – the raw materials they need to manufacture their food – is diffuse. CO_2 exists everywhere and because it diffuses very easily, because it is a gas, its concentration in the atmosphere generally is the same around the globe. Light, too, is everywhere, although its intensity varies depending on shading, clouds, and the time of day. Finally, all minerals and water are contained in almost all soils, at least in concentrations exploitable by plant species that have adapted to local conditions. Therefore, plants do not need to move. Wherever a new plant is established, there is at least a minimal amount of water, minerals, and light, as well as a steady supply of CO_2. Although resources might be limited in absolute quantities or fluctuate with the seasons, a minimum supply is guaranteed. Of course, immobility does present some challenges to plants, but plants' response to these has been successful.

Nutritional self-sufficiency and immobility dictate the basic plan for the construction of the plant body

This issue may be put aside for now so that we can elaborate further on the impact of the type of plant "food" and appreciate the basic design of the plant body. Part of the raw materials (light and CO_2) enter the plant through its leaves and are used in photosynthesis. Common experience shows that leaves have an extensive surface in relation to their volume. A leaf might be small, larger, or much larger, but it is always thin. Why are leaves thin and lamellar? This need is related to light absorption by chlorophylls, through which photosynthesis starts. Imagine a leaf's interior. It is made up of successive layers of cells, each one containing several chloroplasts. Each chloroplast masks the one beneath it, in the same way cell layers do. This scheme of things means most of the available light is absorbed by the top layer of photosynthetic cells, and the layer below can exploit less light, that is, what was not absorbed by the top layer. Therefore, light intensity gradually decreases within the leaf; the deeper it goes, the less intense is the light so that it is not sufficient at the deeper layers. This means there is no point in the leaf being thicker. The optimal construction for a leaf would have only one layer of cells with only one layer of chloroplasts within each cell; however, this leaf would be extremely fragile. So most leaves are 200 to 500 μm (i.e., 0.2–0.5 mm) thick, which is a compromise

Why do leaves have a large surface area in comparison to their volume?

Why are leaves thin?

Why aren't leaves thinner?

between their need for effective light absorption by all layers and their need for a robust construction that does not break easily. The thicker the leaves, the less the chlorophyll they need to contain, so the gradient of light within them may be less steep.

A photoselective antenna cannot be thick and it should have a large surface area to collect maximum light. This is how the lamellar form of leaves emerges; however, the surface may expand only to the extent that it can avoid the risk of breakage. A large surface in relation to volume also increases the probability of CO_2 absorption, which is particularly useful given that its concentration in the atmosphere is a mere 0.037 %. In other words, not only is plant "food" (carbon) diffuse, it is found in limited quantities, so the largest possible absorption area is favoured.

The same constructional needs dictate the structure of photovoltaic antennae

From a statics point of view, a huge leaf is impractical in the same way a huge photovoltaic panel is not practical. The obvious solution is an increased number of leaves. There are plants with a few large leaves and others with many small ones. However, when many leaves exist, there are problems with co-ordination, because each leaf operates as an independent photoselective unit. Leaves are autotrophic (photosynthetic) parts of the plant. Other plant parts depend on the food manufactured by leaves. This food must be appropriately distributed to the rest of the plant, giving priority to the heterotrophic organs in greatest need at any given time. "Any given time" means that relative needs change depending on the season and the developmental program of each plant. For example, before the adverse season of the year for the plant, the photosynthetic product needs to find its way to the plant's storing tissues, usually located in voluminous plant organs such as the stem or root. This stored matter is released gradually during the adverse season so as to maintain the plant, which might be in a state of partial "hibernation." When the next favourable period arrives, this

In plants, the direction of flow within vessels may change, depending on the season or the plant's developmental stage

material is released quickly to support new growth. In another season, though, what needs to be supported is flowering and fruit ripening. Imagine a big tree with several thousand photosynthesising leaves, several hundred storage points, and thousands of developing flowers; add to that the changing direction of transport of photosynthetic products according to seasonal needs. All these processes are coordinated successfully, even if the system appears chaotic at first sight. The various parts of a plant enjoy significant functional autonomy (by planting a twig, one can create a whole new plant) but communicate with one another using chemical signals sent through a network of vessels or from cell to cell so that information about their state is disseminated to all plant parts and behaviour is coordinated. This point is also discussed in other parts of the book.

A few large leaves

How big can a leaf be? The floating leaves of the *Victoria regia* water lily have a diameter of around 1.5 m and a total area of 2 m^2. They are so large, one could almost sleep on them. However, these leaves do not face significant problems with static strength because they float and do not have to deal with wind pressure. On the other hand, an adult oak may have up to 200,000 leaves. If spread on the ground, they may cover an area of about 500 m^2.

Numerous small leaves

The same tendency towards an increased area in proportion to volume is found in roots, which have the task of taking up water and minerals. However, even these resources usually are in short supply in the soil, which is why roots tend to increase in length and branch out to form a dense network. How deep a root can reach depends on the plant, the moisture content of the soil, and the depth of the permanent aquifer. Some desert plants have roots up to 20 m long to reach permanent water reservoirs. The part of the plant above ground might not always be striking, but the underground part is quite voluminous. Of course, a 20-m depth is not impressive compared with the total length of a plant's

How deep can a root go?

root system, that is, the length measured if one places in line not only the main root branches but all tiny rootlets as well. A humble annual clover plant with around 20 small leaves, growing in well-watered soil, needs so much water that although its main root is hardly longer than 20 cm, if all the branching rootlets are included, the total length would be close to 7 m. One may well imagine how long a tree's root system would have to be: for a modest total leaf surface area of 25 m^2, the root length would be tens of kilometres.

What can the total length of the root system be?

In summary, a plant always tends to develop large light-collecting surfaces above ground and correspondingly large water-collecting surfaces underground. This tendency means that a plant's surface area is large in proportion to its volume, allowing it to collect its diffuse "food" successfully. On the contrary, animals, which receive their food in condensed and "packed" form, tend to have a lower area/volume ratio. Furthermore, because animals have to move to find their food, they need an aerodynamically (or hydrodynamically) streamlined body with a suitably positioned centre of gravity. As a result, animal bodies are symmetric along the sagittal plane (*Bilateria*). The need for symmetry in plants, which are rooted on a spot, is exceptionally low and it is rare for some sort of symmetry axis or level to be discerned. A plant attempting to run would present a truly comic sight before collapsing to the ground.

Why are plants usually nonsymmetric?

Plants' lack of limitation in developing nonsymmetric bodies is significant for their survival and is always in relation to their stationary character and manner of nutrition. Their food might well be diffuse, but – with the exception of CO_2, which is found in almost the same concentrations around the globe – water, light, and soil minerals generally are distributed unevenly within a plant's territory. With regard to light, the lack of symmetry is self-evident. Many people remember experiments in school in which they planted lentils on a window sill and watched them turn to the light and

Bodies are nonsymmetric because of nonsymmetric resource distribution

develop nonsymmetrically. This also may happen in the soil if resources are distributed nonsymmetrically. For example, buried animal excrement is an exquisite delicacy for roots because it contains high concentrations of nitrates and phosphates. Roots chemically detect the location of nutrients and orient their growth – by appropriately directing their rootlets – towards them, thus becoming asymmetric. The same thing happens when roots encounter an obstacle on their way to resources; they make a detour just like people would walk around a fallen tree trunk blocking their way.

Immobility in plants means fixation at the point of their establishment

This probably is a fitting point to further elaborate on the issue of plants' lack of motion. Undoubtedly, plants are immobile when viewed in human terms; however, this immobility should be considered only in terms of the plant being fixed at its establishment point, because turning towards the light is not a movement by the whole plant but only by some of its parts. The same is true for a root turning towards resources or moving to avoid obstacles. These movements are so slow, compared with those to which we are accustomed, that we hardly notice them. On the contrary, some other movements – for example, the defensive shying of *Mimosa pudica* (touch-me-not) or the active trapping of insectivorous plants – have become famous because their speed is comparable with that of animal movements. These rapid movements are studied in Chapter 9 of this book.

The movements of a stationary organism

There is no constant size in plants…

…because they keep developing throughout their lives

Adult animals are relatively stable in size. A human's height is around 1.75 m, with a few minor deviations from the mean value. Major deviations are not pathologic but are classified among noteworthy exceptions, and often they create – sooner or later – problems in the daily lives and health of such individuals. A person usually reaches mean height at adulthood and stops growing after that. However, this is not true for plants, which keep growing throughout their lives. Not only do they keep growing, but the size of two individuals of the same age might differ significantly. A pine tree established in a position with limited resources becomes a dwarf, while a sibling established in a more favourable location may turn into a giant during the same period. Plants literally spread as far as they are allowed to grow.

Among humans, poor – as opposed to satisfactory – nutrition in childhood might lead to a difference of a few centimetres in height. In trees, however, this difference might be tens of metres. In other words, plants are characterised by plasticity not only with regards to their final size but also in terms of growth rate, which is determined to a much greater extent by environmental conditions. This plasticity is not limited to growth rate or body size, but extends to the proportions of a plant's parts. For example, in animals, movement imposes not only a streamlined shape and overall symmetry, but also a stable ratio among body part sizes. When one's arms grow, the legs grow proportionately so that their length ratio remains stable. Imagine the difficulties of a person unfortunate enough to have legs and a torso a mere 10 % above average while the arms are only 10 % shorter. He or she could not perform daily tasks, such as looking after personal hygiene. The opposite is true of plants. For example, the weight ratio of aboveground and underground parts is not stable and depends greatly on environmental resources. Assume that in a clover plant growing in well-watered soil, the weight of its aboveground parts (leaves and stem) equals that of the underground parts (roots), that is, the weight ratio is 1:1. If watering is limited or there is no rain, the water available in the soil and absorbed by the root cannot replace what is lost through the leaves. A solution to provide a new balance and prevent the plant from withering may be for the root system to continue growing while the leaf surface area is suspended; this happens as the plant may lose some of its leaves. The new balance depends on the extent of drought stress. After attaining the new balance, the weight of the foliage might be one tenth that of the root system. Actually, plants keep changing their foliage/

In plants, there is no steady part ratio...

...because the ratio changes with environmental conditions

Foliage-to-root weight ratio depends on light and water availability root ratio, particularly those established in seasonally changing environments. Many evergreen Mediterranean bushes drop a significant part of their foliage, sometimes as much as 50 %, to survive the summer drought, replenishing it in the autumn when the rain returns.

Therefore, a plant's growth and development program does not stop when it reaches adulthood; it remains active throughout the plant's life. Furthermore, the program keeps adapting and changing depending on environmental pressure and resource sufficiency, that is, depending on the season. Indeed, the program might be expressed unilaterally: plants do not have a stable shape or specific symmetry in time and space; they keep changing their form with the seasons and they turn in this or that direction. In other words, their form depends on their site of establishment.

What is (genotype) and what appears to be (phenotype)

Students often are told that all an organism's genetic information comprises its genotype (which generally is stable) whereas the phenotype includes all the organism's properties, including its form. Genotype is determined by one's parents, from whom genetic makeup is inherited. The same genotype may result in different phenotypes in different environments. The range of phenotypes that might result from the same genotype depends on the organism, which may be characterised as having higher or lower "phenotypic plasticity." The range of phenotypic plasticity among plants (as in all sedentary, immobile organisms) is wide, whereas that of animals is narrow. Bears of a specific species hardly differ from one another in terms of form, size, or body shape. However, in trying to locate two similar pine trees in the woods, one would discover they simply do not exist. In other words, among plants, the same genotype – that is, the same genetic makeup and information – may result in a variety of sizes, shapes, and forms within a species. The phenotypic plasticity of plants also is displayed in their biochemistry. For example, a warm-blooded animal, which

Why are there no two identical trees in a forest?

Phenotypic plasticity: when the same genetic information produces infinite forms

keeps its internal temperature stable, has enzymes designed to act optimally within the narrow range of its internal temperature. A plant, whose body temperature ranges widely depending on the temperature of the environment and is at the mercy of changing weather conditions at its location, cannot have one enzyme for all seasons. Rather, a plant modifies its enzymes depending on the prevailing external temperature levels or synthesises new enzymes when the temperature changes. In other words, a plant's internal makeup has a phenotypic plasticity comparable with that of its form.

Plants' nutritional self-sufficiency and diet, therefore, determine their way of life; it is a life that does not require movement, but it affords the possibility of continued growth to sedentary individuals in a fixed position. Growth takes place in stages: it is fast-paced during the favourable period and becomes slower or stops during the unfavourable period. It also turns in the direction where there are abundant resources and finally shapes organisms with no set size or symmetry, with significant plasticity with regard to form, chemistry, and behaviour. However, the immobility of sedentary organisms also raises a series of other questions: How does a sedentary organism reproduce? What is sex like among these organisms? How does a plant find its mate? Because it cannot run away, how does a plant face its enemies? Why does it not become extinct when devoured by herbivores or as a result of pathogen microbes? How did natural selection respond to these challenges? Because the answers are so intriguing, all these topics are investigated in detail later.

The Concept of Death in the Plant Kingdom

A particular way of life requires a particular way of death. Indeed, the concept of death in the plant kingdom differs at the population as well as the individual level. If you shoot all the rabbits on an island, you will render the species extinct unless you bring a couple of rabbits from another place. If you felled all the poplar trees, in a few years there would be a new wood. This makes sense because we know that new shoots would come up from the roots. However, what if the roots were dug up and burned? Again – over a longer period, of course – a new poplar wood would grow from the seed bank patiently waiting in the soil. Seeds, the products of sexual reproduction of adult plants, contain all the genetic information to form a new plant, just like the fertilised ovum of any animal. In contrast to the fertilised animal ovum, however, plant seeds have certain major comparative advantages: they need not stay within a uterus to grow and develop, nor do they need hatching or parental care. Their main advantage is their durability over time and through environmental adversity. Seeds can survive darkness, frost, high temperatures, and sometimes even fire, drought, or floods to bring forth new plants when better days come, even if it takes tens or hundreds of years. No seed bank like this exists for animal "seed." Animal reproduction demands the simultaneous presence of both parents for a short or long period. As is shown in Chapter 4, the exceptional resistance of plant seeds to time and adversities is one of the many reasons plants did not suffer – as animals did – from mass extinctions in the geologic past.

Embryos in plants do not need extensive gestation or parental care

Plant seed banks

The aforementioned observations about death concern plant populations, but how do they apply to individual plants? There are not many apparent differences in death among plants, except that trees die standing and some plants may live for several thousand years. The reasons for this unusual longevity are examined in Chapter 3. It will also be shown that, at least in some cases, the final cause of death is incidental. If these plants do not fall victim to fire, human intervention, or an unusual attack by pathogens, they may be considered virtually immortal. However, plants constantly are faced with (or more precisely, often impose on themselves) the ageing and death of some of their parts. If you cut off an animal's leg, it might survive the haemorrhage and infections, but it would have a tough life and come to a bad end because it could not face competition or predators. If one's leg or foot is amputated, medical care and social security would allow an acceptable level of quality of life and longevity comparable with that of fully able-bodied individuals, provided he or she did not try to walk the streets of Athens. Of course, in the case of heart or kidney failure, only transplantation could help the patient, and with only dubious results. Unfortunately, losses suffered by animal bodies cannot be replaced; organs are produced only once in a lifetime.

Partial death and rejection of organs…

Almost daily, plants face losses they soon replace. These losses are attributable to consumers or to programmed ageing and dropping of selected organs during specific seasons. The oak tree mentioned earlier may have up to 200,000 leaves, which, because the tree is deciduous, will drop in the autumn because, in low winter temperatures, leaves are not photosynthetically efficient. These leaves will be replaced the following spring in an equally large number. If the average lifespan of such a tree is 200 years, then it will produce a total of 40 million leaves. If these leaves were spread on the ground, they would cover more than 70,000 m^2. The same tree also produces hundreds of flowers (i.e., genitalia) every year and, correspondingly, thousands of seeds to be stored in the soil bank. There is nothing similar in the animal kingdom. In some herbs, the entire aboveground part of the plant dies during the unfavourable period of the year, whereas the underground part is maintained to bring forth a full plant the next favourable season. Biotechnologists would never – even in their wildest dreams – conceive of the contingency of such immortality in humans.

…which are replaced in the next growth period

How do plants manage to replace their buffeted, consumed, or voluntarily aborted organs? Here lies probably the ultimate major difference between animals and plants – the so-called meristematic cells located in the meristem of plants. These cells are undifferentiated, that is, no developmental processes have been initiated within them to produce their ultimate shape and size or function. In other words, meristematic cells are nonspecialized, in contrast to the rest, which are specialized. Examples of specialized cells are the epidermal cells protecting the body of the plant; the photosynthetic green cells within leaves; the root hairs, that is, the cells of the root epidermis, which are specialised in absorbing water and minerals; the vessel-forming cells (tubes) for transporting substances within the plant body; the colourful petal cells in flowers; the juicy cells of fruit; and so on. All specialised cells, with their different structure, composition, and function, originate from meristematic cells. In other words, meristematic cells are totipotent cells, which – contrary to other cells – are capable of propagating with cell division and of giving birth to specialised cells while keeping their own population intact. Because of this population of cells that maintain their totipotence, juvenility, and division potential, the plant may forever renew the organs it loses and may increase its size throughout its life.

Nonspecialised, juvenile, and totipotent cells everywhere...

...which can replace damaged or rejected organs at any given time...

Meristematic cells are found in zones dispersed throughout the entire plant body. Some are visible to the naked eye as small bulges at the tips of stems and branches or in leaf axils, where the leaf pedicle joins the plant stem. Depending on their position, these "eyes" will bring forth flowers, leaves, or new shoots. The signal for differentiation to begin so that various organs are produced comes from the environment. The plant perceives information from the outside world, such as the duration of night and day, prevailing temperature, or existence of competitive plants nearby; processes it internally to confirm its credibility and significance; and translates it into chemical messages

...when they receive corresponding environmental stimuli

to the meristematic cells, signaling them to divide and differentiate to produce the corresponding organ. The terms *perceive*, *internally processed to confirm credibility*, and so on should not sound strange. These processes actually occur, although one may believe that decision making on the basis of perceived stimuli is an ability only animals possess. To elaborate, meristematic cells of the aboveground plant parts, depending on their position and the season, may produce new leaves, flowers, or branches. Suppose a branch bearing these meristematic cells were cut down and planted in the soil so that these cells were buried under the surface. The buried part of the branch would bring forth not new leaves or flowers, but roots. For this destination change to occur, the environmental signal (which is now totally different because the branch is not in the light but in darkness) must be processed to lead to a new "decision," allowing the same undifferentiated, juvenile meristematic cells to produce a new organ with a different function.

Juvenile tissues are dispersed and invisible throughout the interior of a plant body, because, as was mentioned earlier, the whole plant grows and develops during its entire life. Tree trunks, for example, increase their diameter continuously as a result of the action of the so-called secondary meristem, which, like an internal cylinder, runs along the full length of the trunk. The cells of the secondary meristem annually renew the plant's vascular system by adding new vessels (tubes) so that water may be transported upwards and minerals may circulate. As is shown in Chapter 3, plants need not worry about obstructed arteries, nor do they require surgery to unblock them. They are capable of spontaneous bypass, as a short-term solution, until the juvenile cells of the secondary meristem produce new, fresh, and healthy vessels. This process is repeated every year, and

In trees, vessels are renewed annually...

the new vessels form concentric circles from the trunk centre to its periphery. This makes it possible for us to determine a tree's age merely by counting the annual activity of the secondary meristem, which is visible to the naked eye on the sections of cut trees. These are the annual rings of wood. A second cylinder of juvenile cells, concentric to the one producing new vessels but lying more peripherally, is responsible for replacing the worn cells of the tree bark. This structure is the so-called cork cambium. The product of cambium activity in some trees is commercially valuable for manufacturing cork.

...giving us the opportunity to determine the age of a plant from its annual rings

In other words, the dispersion of meristematic cells throughout the plant body during its entire life warrants continuous growth and smooth renewal of organs lost to consumers or dropped, to ensure better function of the remaining plant parts. The activity of meristematic zones might be nonsymmetric, which explains the temporary changes in plants' growth direction, as well as their powerful phenotypic plasticity.

Is there anything similar in the animal kingdom? It is not difficult to comprehend that if something like this existed, at the same quantity and quality level, then the concept of physical disability generally would be unknown and humans would not worry about damaged vessels or degeneration of mental capacity. However, blood cells (red and white cells, platelets) are produced continuously throughout our lives, so those that are inevitably damaged may be replaced. The tissue responsible for this task is the bone marrow. Here, juvenile cells, the so-called stem cells, maintain their capacity to divide so that they produce blood cells while maintaining their own numbers. A second group of juvenile cells are dispersed in the areas under the skin and the interior epithelia (the internal lining covering the lungs and digestive system) and renew the damaged cells throughout life. Other than these cells' activities,

The concept of physical disability is unknown among plants

Corresponding nonspecialised juvenile cells among animals (the so-called stem cells) are capable only of patching repair work

however, there is no possibility in the human body for continuous, mass cell replacement. *Mass* is the key word here, because recent decades have revealed that a limited number of organs have small stem cell groups that undertake small-scale repair work. However, there is no potential for full replacement, only a little patching here and there, nothing like what exists in plants.

Research into human stem cells has attracted a great deal of funding in recent years and has made even more promises. However, the optimism and urgency may be unwarranted. So far, practical results are both scant and controversial. The effort mainly concerns the isolation and conservation of stem cells (e.g., from the liver) of healthy individuals to possibly be used in the future by the same people to regenerate the corresponding organ in case of an accident or disease. This process functions as a stem cell bank. Because the bank's reserves consist of donations that will return to the same donees, the problem of rejection is bypassed. Furthermore, the reserves are obtained from specific organs containing stem cells that can be isolated and conserved. Another line of research concerns the reprogramming of stem cells so that they may be used to regenerate other organs, not only those from which they were isolated. However, this has proven even more difficult because the stem cells found in various organs have lost their totipotence: when they divide, they produce cells with functional properties corresponding to the organ of their origin. Only embryonic stem cells at the early stages of development are totipotent, which means they can produce cells corresponding to the receiving organ, that is, to the organ to which the cells are transferred, because they are programmed to differentiate as soon as they are within the internal environment of an organ. However, because the isolation of stem cells means the death of the embryo, research into totipotent embryonic stem cells has been forbidden. It is not difficult to imagine the crimes that might be committed with the practical application of embryonic stem cells, who might exploit such an application, and who might fall victim to it.

Given this state of affairs, one can summarise the similarities and differences between animals and plants with regard to replacing damaged organs. The basic similarity is that, in both cases, adults maintain some parts of their system in a juvenile state; these are undifferentiated cells tasked with replacing lost cells throughout the organism's life. In animals, these juvenile cells are called stem cells and in plants, meristematic cells. Their basic cellular function is the same; however, in plants these cells are far more numerous, their function extends to replacing whole organs, and their reprogramming is easy, occurs daily, and responds to distinct environmental signals.

As has already been discussed, reprogramming of animal stem cells ranges from difficult to unfeasible, whereas that of plant meristematic cells is easy and essential for daily life. Consider, for example, an immobile plant fixed in a specific location: ground subsidence might reveal part of the plant's root system, whereas a landfill might cover part of its shoot and corresponding leaves. In this position, the root is useless for absorbing water and minerals and the shoot incapable of photosynthesis. These environmental changes, however, leave the plant indifferent, as the root now exposed can reprogramme its "eyes" so that instead of giving forth root branches, as it normally would do, it now produces shoots with leaves. Similarly, the shoot – now buried – may bring forth roots with water-collecting rootlets instead of shoots with photosynthetic antennae (i.e., leaves). This flexibility and adaptability are absolutely necessary for an organism that cannot respond by fleeing.

Roots often are revealed and aboveground parts buried at some point in a plant's life

Problem? What problem? The reprogramming of juvenile cells will soon restore the situation.

Moreover, plants have gone a step further. Not only undifferentiated meristematic cells, but also differentiated and specifically specialised somatic cells may – under certain circumstances – return to their embryonic

Reprogramming also concerns specialised, somatic cells

totipotent state. Thus, these cells may begin dividing again, ultimately regenerating a whole, healthy plant from a single somatic cell. A milestone in the corresponding field of cell culture was the laboratory production of carrot plants from differentiated root cells in 1958. Such plants may be transferred – following the appropriate, gradual acclimatization – from the laboratory to the garden and thrive. Combining cell culture with modern molecular biotechnology methods is easier (and relatively cheaper) in the case of plants and already has produced practical results, or – to be more precise – the promise of this research field are much more realistic.

The first carrot plant made from specialised root somatic cells was produced in 1958 and the earth did not move

Returning somatic animal cells to their embryonic state and using them to produce clones is exceptionally complex, time consuming, and expensive. In 1996, after hundreds of failed attempts, this method resulted in Dolly, the congenial ewe that survived for a few years with the help of medicines. She was an ailing, delicate animal, rather unlikely to survive in the real world. It would have been impossible for Dolly to make it in a flock of sheep. Her early death disappointed the conceited individuals who believed the time had come for them to acquire – with the help of a great deal of money and the necessary laboratory support – clones of their own beauty and intelligence. However, a case may be made that technology will advance in the future, with continuous improvements in cloned products. Maybe future Dollies (or one's own clone offspring) will enjoy better health than the average individual. Perhaps instead of suffering from early ageing (like Dolly, who was euthanized at age 6 instead of dying peacefully at 13), they will live longer. This possibility is being claimed by many biotechnologists. However, the reader, whether or not an expert in the field, should be aware of how Dolly was created. The method, called "somatic cell nuclear transfer," includes the following steps: First, a mature cell is isolated microsurgically

Dolly, a similar animal achievement, became a TV celebrity in 1996. She was an ailing, delicate animal whose life was supported by medication; Dolly met an early death.

from the breast of the donor sheep; then, the cell's nucleus is removed and reintroduced (again, microsurgically) into a nonfertilised (also isolated) ovum, the normal nucleus of which has been removed. The hybrid cell is then stimulated with an electric current so it may start dividing. The mass of cells produced are implanted into the uterus of a surrogate mother. One wonders how it is possible, after such hardship and tribulation, for the organism produced to be better than one produced conventionally through sexual intercourse, which is more pleasant and less expensive.

Although carrot (and several other plant) clones produced in the laboratory manage much better than Dolly, this book does not address the practical applications of plant cultivation. Therefore, let us ask if somatic cells turn into embryonic ones in nature, and why. These questions obviously relate mainly to reproduction, because an embryonic cell produced from a somatic cell can divide and develop under suitable conditions, and produce a new, complete organism. Because such an organism is genetically identical to its donor – and there is no mixing of parental genetic material, as in sexual reproduction – it is, in effect, a clone. Are there clone plants or animals in nature? The answer is unequivocal: Natural cloning is nonexistent among animals (except for corals and other sedentary animals) but occurs daily in the plant world; 40 % of plants create natural clones because they can reproduce asexually (without the presence of two progenitors). Later (see Chapter 5), this book will show that this exceptional capacity also is associated with immobility, because it allows reproduction even if a plant cannot approach its mate. For example, if only a single fig tree were left on Earth, its reproduction would not be impossible. If the tree's branches, or a even single vine, were cut and planted, a cultivation of clones would be

Plant cloning is a daily phenomenon in nature...

...and contributes to the reproduction of individuals and the resistance and longevity of populations

created. In other words, plants' inability to approach one another does not prevent them from reproducing; this attribute is related to the resilience of plant populations over time and their avoidance of mass extinction, which is common in the animal kingdom. Plants' cloning potential relates not only to the totipotence of their meristem, but also to the somewhat rarer but extant dedifferentiation of somatic cells into embryonic ones. This dedifferentiation is not encountered among all plants; however, it is common to see pictures of the mature, succulent leaves of certain *Crassulaceae* species (e.g., *Kalanchoe daigremontiana*, *Bryophyllum tubiflorum*) at a certain time of the year bearing entire miniscule plants with roots, stems, and leaves. The new plants come from mature leaf cells that regress to their embryonic state and produce plant clones on the leaf surface. The clones detach from the maternal plant, then grow and develop on their own.

Plants as Modular Organisms with Increased Autonomy of their Parts

Plants have an elaborate circulatory system

What is the equivalent of the circulatory system in plants? Do they need one? Plants have two types of vessels: the so-called xylem vessels, responsible for transporting water and minerals absorbed by the root from the soil to the aboveground parts, and the phloem vessels, which transport photosynthetic products – plant food – from where they are manufactured (the leaves) to where they are consumed (the rest of the plant) or stored (mainly the roots and stems). In the xylem vessels, the movement is against gravity, that is, from the roots to the top of the plant. In the phloem vessels, the movement may flow in any direction, depending on the relevant position of leaves and consumption or storage points. A developing fruit or flower may lie above the leaf providing its food, whereas the root is always below it. Movement through vessels needs to be driven by a force, a kind of pump. However, plants have no pumping organ equivalent to the animal heart, which is a fundamental difference. In Chapter 3, this book demonstrates in some detail that although there is no central pump, there are many small pumps operating within the plant, using either force or suction. To transport water within the xylem vessels, each root functions as a small forced pump and each leaf as a small suction pump. However, pumping has a high energy cost. For water to be drawn from a well, it needs the energy provided by a pump. A special feature of plants is their excellent energy economy when pumping water. Therefore, in the forced root pump, the energy inevitably consumed by the root cells to absorb minerals from the soil solution (where they are found in low concentrations) and to deliver them to the base of the xylem vessels (where their concentrations are higher) is also used indirectly to supply the vessels with water. The operation of the suction pump of the leaves entails no energy cost for the plant, however absurd this may seem. In the case of plants, the physical and chemical properties of water are used optimally in combination with the optimal diameter of xylem vessels assigned this task; it is as if the water "goes up on its own." As a whole, therefore, the water's upward movement through the xylem vessels has no cost, even though it flows against gravity. However, the water's movement (along with the dissolved sugars) in phloem vessels has some energy cost, yet this is disproportionately low relative to the quantities transported. The mechanisms and driving forces drawing the water upwards are rather easy to comprehend. They are referred to in the next chapter, when the book examines the maximum height to which water can ascend, at no cost, within the plant body. This height also determines the maximum tree height, which is around 120 m, an impressive size

However, they lack a central pump

They manage equally well, though, with a system of many small forced and suction pumps dispersed throughout the plant body

indeed. The corresponding movement in phloem vessels must be explained in more detail; however, that topic is beyond the scope of this book.

Plants also lack a central behaviour-coordinating organ and localised sensory organs

Plants lack not only a central pump, but also a central coordinating system; they have nothing equivalent to a brain. However, plant development and behaviour are anything but uncoordinated. Like animals, plants perceive the environment, light – in terms of both intensity and quality – touch, temperature, humidity, their neighbours, and their temporal and spatial position; they measure time; they perceive chemicals in the atmosphere; and they use a chemical language to communicate with their neighbours, pathogens, or symbiotic bacteria and fungi. The information plants receive from the environment is processed and gauged so they can regulate their development and behaviour in an intelligent manner, as animals do; they would have been doomed to extinction if they did not. The reader can find scattered information throughout this book about the way plants perceive their environment without apparent sensory organs and about how they process this information without an apparent central coordinating organ and nervous system. As in the case of the circulatory system – with its many small and partially independent pumps instead of a central one – when a plant perceives its environment, stimuli are processed in numerous scattered and partially independent centres throughout the plant. However, the question regarding plant intelligence is elaborated on further at the end of the book. For now, it is enough to keep in mind that plants, as compared with animals, are characterised by a logic and philosophy of decentralisation of their functions and relative independence of their parts, at both the organ and cellular levels.

However, they perceive and measure environmental signals and react in an exceptionally coordinated manner

Central power and decentralisation

Chapter 3
Why Trees are Almost Immortal and Other Related Issues

Life span limitations

What is it that determines an organism's life span? If the organism survives the initial embryonic and infant stages, when it is at its most vulnerable, it very likely will reach reproduction age. In organisms of the animal kingdom, this is the point at which growth and development normally cease. If the organism's inherent defensive capacity is satisfactory, it can tackle pathogenic microbes successfully. Furthermore, if it escapes hunger and, with a bit of luck, is not eaten by its predators, it might reach the age of maturity. After that, the countdown starts with a gradual decline in physical and mental functions, whereupon the organism becomes as vulnerable as it was at the early stages of life. In nature, the life span an animal achieves depends on skill and circumstances, which, in the best of cases, might extend life to the limits characteristic for its species. However, what about the limits for plant life?

Plants that flower only once and then die

For some species, called semelparous, life ends after a single flowering and fruit-producing period. Before it flowers, the plant develops without blooming; this period might last from a few months (annual plants) to several decades. An extreme example is agave (*Agave americana*, a native succulent of Central America acclimatized to many parts of the world), which has a single rich blossom after 50 to 100 years of life without love. As if exhausted by this overwhelming and unique experience, the plant collapses within a few weeks and becomes a desiccated brown cellulose skeleton. The final act of this drama is the dispersal of its seeds, the miniscule descendants that will seek their fortune in the struggle for survival and reproduction.

Annual Rings as Climatic Memory

In ligneous (i.e., woody) plants and trees in particular, maximum life span limits are rather vague compared with those among animals. Humans, even under excellent hygiene conditions and having all possible medical interventions, cannot live more than around 90 to 100 years. As for plants, it is well known that some (e.g., the plane tree) may live for centuries. To determine which plant species are the champions, one must know how to determine the age of a tree accurately.

Some trees live for centuries

When a tree is cut, the surface of the stump reveals a series of concentric circles, the so-called annual rings. Each ring corresponds to 1 year and is the result of climate seasonality. Almost everywhere on Earth during a calendar year, there is a favourable and an adverse growth period for plant development. During the favourable period, the plant grows, whereas in the adverse period, its normal functions are curbed and in some extreme cases the plants go into proper hibernation. For example, consider a conifer at a high altitude in a temperate zone (i.e., in a region of medium latitude). Obviously, the favourable season is summer, when mild temperatures and water availability make growth possible. On the contrary, winter is the adverse season, because temperatures are low and water may be in its solid state in the form of ice, that is, not available for plant use. Consequently, and depending on the prevailing microclimate of the area, growth starts, say, at the end of spring. During this period, the meristematic (embryonic) layer of the cambium, directly under the living bark of the trunk and branches, starts producing cells that will become the new plant vessels, replacing the old ones lying nearer the centre of the trunk, which are already inactive. Following a brief growth period, these new cells enter their last developmental stage, which is programmed cell death. During this stage, all cellular organelles (mitochondria, nucleus, plastids, cytoskeleton, etc.) decompose. At the same time, the cell wall is lignified while at the upper and lower part of the cells, communication channels are formed with their bordering cells above and below. Under an electronic microscope, the contact

What are the annual rings in tree trunks?

Trees renew their vessels once a year, because the old ones have suffered damage

surfaces between two successive cells look like grids. Along the height of the plant, the successive – already dead – cells form ligneous (xylem) vessels that run along the trunk from its base to its top. All the adjacent tubes along the periphery of the trunk form a cylinder. These tubes function as vessels transporting water from the base to the top, from the soil, through the roots to the leaves, and finally to the atmosphere.

This process is repeated throughout the favourable period, adding new vessels in a centrifugal arrangement. When the next adverse period starts (e.g., when autumn starts and the first cold temperatures arrive), growth is curbed gradually and the vessels produced become successively smaller until their production finally stops. The difference in the size of the vessels at the end of a period is visible to the naked eye as a darker zone, because it reflects less of the light falling on it. Production of the annual ring is now complete. During the adverse period, the vessels are functionally inactive, at least with regard to water transport. This role will be undertaken by the vessels of the next ring the following summer. This is how, throughout their lives, trees add a ligneous layer around their trunk. The easy distinction between small vessels at the end of the period and larger ones in the beginning of the next makes it possible to count the rings and, therefore, calculate the tree's exact age. Of course, it is not necessary to cut a tree down to determine its age. One can merely bore out a small "carrot" from the periphery to the centre of the trunk using a special instrument that causes no significant harm to the plant.

However, the vessels of the previous year – although inactive – remain in the trunk

Counting successive vessels, from the oldest lying centrally to the youngest at the periphery, indicates the age of the tree. The first ring is the tree's birth certificate.

This method of determining a tree's age is known as dendrochronology, and some of its important applications are presented later. At this point, however, the following questions might be raised: What is the purpose served by the continuous production of new xylem and the functional redundancy of the old one? How does the cost of production of previous rings pay off? Is there an adaptive significance to the phenomenon? Could one just say it is a necessary evil? Old vessels may (or may not) suffer wear and tear, and because they are made up of dead cells, repair is impossible. On the other hand, the tree grows annually, accumulating both volume and weight. Therefore, its needs for support increase, and these requirements are met by the continuous deposition of new wood.

Why do we keep seeking trees with a longer life span? Possibly out of sheer curiosity. To give a personal example, if I notice a cut tree trunk in the woods, after overcoming my initial aversion to the unjustifiable and most likely violent death of the tree, I usually become engrossed in studying its rings. It is fascinating to count the lines reflecting the tree's efforts to survive through decades. It is also an opportunity for me to look back and review my own life events. The 16th and 27th rings correspond to the birth of my children and the 46th to my entering university, which gave me the opportunity to learn how to assess these rings. Other important points in time are reflected in the rings, such as birth and death dates of my favourite poets and historical events that had a great impact on my country.

Do natural phenomena also leave their trace on the rings? Can one discern the big drought of 1997 or the tragic, harsh winter of 1941?

How the climate is reflected in the annual rings

In the example given previously of the alpine conifer, one may assume that the width of its rings depends on climatic conditions at any given time: the more favourable the environmental parameters affecting growth, the wider the ring. The alpine climate is particularly variable, and the length of the period favourable for growth may fluctuate significantly from one year to the next. The main factors affecting ring width (i.e., plant growth) is water (i.e., rainfall) and temperature, both of which have a positive impact provided the temperature does not exceed the optimal value. The interaction between temperature and water supply in determining the growth rate observes the law of restricting factors. This is a simple law that – in its broader sense – may be applied from physics to everyday life.

The law of limiting factors in the development of plants and daily life

For example, until relatively recently, the mean speed from Corinth to Athens was restricted by the well-known traffic jams at Kakia Skala, where the road was narrow. There was no point in widening the rest of the road or driving faster to Kakia Skala. The only way to increase one's mean speed was to expand the bottleneck point, that is, to eliminate the restricting factor. Once this part of the road was widened, the mean speed increased but was now restricted by the speed limits indicated by the signs. Although this restriction is of course only theoretical, let us accept it for the sake of argument. If the traffic police decided to eliminate the speed limits, the mean speed would increase again to the point at which it might be restricted by the limits on a car's motor inherent in its manufacture. From this simple example, a definition can be formulated: when the speed of a process depends on more than one factor, the speed is determined by the factor of restricting quantities (or restrictive values). With regard to plants, the growth of an individual may be stunted if it is deprived of water, light, and mineral nutrients from the soil or if the temperature is too high or too low. If there is not enough water, there is no point in fertilising the soil. If, however, the water requirement is met, the growth rate will increase until it is restricted again by the next factor, the supply or value of which would need to be corrected.

The width of annual rings is affected by the temperature and rainfall during the year in which they were produced

With regard to the alpine conifer, a question may be raised whether temperature or humidity is the main restrictive factor for plant growth (hence, the rings). If one selects an individual thriving on a rocky substrate with a steep incline and shallow soil (i.e., where drainage is extreme), the most likely restrictive factor is the lack of water. In this tree, the drought of 1997 would be recorded very clearly in the narrower rings. If, however, one chooses an individual thriving next to a stream of permanent flow, the growth of its rings would be far less sensitive to rainfall amounts. In this case, the narrow rings might be attributed to summers that were too hot or extremely cold, because plant growth would be far less sensitive to water, which is amply provided.

Annual rings and past climatic conditions

At this point, readers might contend that there is no reason to injure a tree to discover there was a drought in 1997 because climate changes, at least for the past 100 years, are recorded in meteorologic archives. However, what about the changes that occurred before the systematic recording of climate parameters? For the distant past, only sporadic data exist, mainly about extreme weather phenomena, recorded by counsellors to noblemen and emperors or by anonymous monks. Indirect descriptions about a good or bad crop might imply corresponding weather phenomena, but these would be of only local significance and might well be based on subjective judgment.

Old-tree hunters

When scientists first became involved in dendrochronology, their attempts to relate ring width to known climate changes may have started as an attractive intellectual game. However, the prospects of getting information about palaeoclimatic conditions turned the attention of many scientists in this direction. One may ask why there is such an interest in the palaeoclimate. The answer is that it can provide us with keys and interpretations regarding the factors that determine climatic change, at least in midterm periods. For example, if periodic climatic phenomena are discovered and properly correlated with other periodic phenomena of a corresponding frequency (e.g., solar spot activity), this improves the capacity for future forecasting. This is why "old-tree hunters" appeared in the middle of the last century. This group was mostly made up of university professors interested in astronomy, the climate, archaeology, plant biology, and palaeo-ecology.

In 1953, driven by rumours, they discovered a population of age-old pines of the *Pinus longaeva* and *Pinus aristata* species in an extensive mountainous region stretching from Colorado to North California. The oldest individuals were found at the highest altitude limit for tree growth, which in that region is between 3,000 and 3,500 m. A short while later, in 1957, Professor Schulman discovered a tree that was 4,723 years old and named it Methuselah; it is still the oldest living organism on Earth. Methuselah was already more than 2,000 years old when the Parthenon was built, more than 1,000 years old when the Minoan civilisation was destroyed, and a young tree of about one or two centuries when *Homo sapiens* learned how to put together the first stone constructs. From 1957 to 2011, it grew by 54 years, a mere 1 % of its lifetime, whereas many humans born in the 1950s are already elderly. I imagine that when Professor Schulman finished counting, he must have counted the rings again and again before he realised what he had in his hands; he must have felt the same as Professor Andronikos upon opening King Philip's tomb.

"Methuselah" of the White Mountains in California was already 2,000 years old when the Parthenon was built

For the sake of comparison, the oldest members of the animal kingdom are the Galapagos Island turtles, which live a mere 150 years. In other words, there is no turtle alive today who saw Charles Darwin disembark on the islands.

Longevity Elixirs

What is the elixir for the longevity of Pinus longaeva?

It is not easy to discern with any degree of certainty the strategies of *Pinus longaeva*'s exceptionally long survival. First, the risk of death by fire is negligible. These trees live at some distance from one another, and because of the lack of low undergrowth and flammable materials on the ground, a random fire (e.g., due to lightning) would not spread easily. Furthermore, there are no known pathogenic microorganisms that attack these pines or insects that live off them. Other parameters contributing to this longevity include low temperatures and an extremely dry environment, which prevent microorganism growth, and a strong chemical defence mechanism based on excretions of thick resin, which functions as an antibiotic. Additionally, the tissues of these trees are made up of densely arranged cells and thick cell walls, rendering them difficult to chew and digest. Therefore, the tree, as well as its particular organs, reach a very old age. For example, the leaves of most other trees remain on the plant from a few months to 4 to 5 years before they finally fall. In the case of *Pinus longaeva,* however, they stay on the tree for about 20 to 30 years. Furthermore, these trees not only die standing but remain standing for several centuries after their death. This is a result of the cold, dry environment that does not favour the decomposition of organic matter as well as to the trees' ligneous structure, which is particularly durable.

Therefore, is an environment unfavourable to microorganisms and a strong defensive system enough to create a Methuselah? That likely is not the case; apparently, a slow growth rate also plays a role. As discussed in Chapter 2, plants do not grow to any specific size and their meristems (i.e., their embryonic tissues), properly arranged within the plant body, enable continuous growth and development. Experience has shown that in the plant kingdom, the longest living organisms are those that do not grow fast. For example, at the age of 3,000 years, an individual of the *Pinus longaeva* species may have reached a medium height of 10 and 15 m and a trunk perimeter of approximately 8 to 10 m. The diameter increases by a mere 0.2 mm per year. A slow growth rate also implies low metabolism and, therefore, a lower probability of errors and less maintenance cost. However, it is not simply a case of

Make haste slowly: in plants, the oldest individuals are those that grow slowly…

. . .or occupy the most inaccessible, poor, and generally unfavourable habitats

making haste slowly. Wherever populations of age-old trees have been studied, the oldest individuals have been found in the less "favourable" microenvironments. Rocks with almost vertical faces, extremely poor soils, windswept peaks, and low temperatures host the eldest trees, indicating that what appears to humans as a tough environment actually results in longevity among plants. This unfavourable climate translates into scarce resources, mainly water and nutrients in the soil, leading to a reduced growth rate, low metabolism, and longer life expectancy. It seems, therefore, that the price plants pay for living longer is a low metabolism. It is not clear whether calculations to this effect have been made, but it would not be surprising if within a plant species, the product of growth rate and life span achieved were stable.

Humble living is the price for longevity

Something similar has been proposed for the life span of mammals, in which the total number of heart beats seems to be a nearly stable value of 10^9, or roughly one billion beats during a lifetime. This is true for mammals thriving in the wild; for example, the life span of a field mouse with 350 pulses per minute is 5 years, whereas that of an elephant with 50 pulses per minute is 35 years. Exceptions to this rule are humans and pets. If the formula were applied to humans, one would conclude that they should live for about 25 years, which probably was the case before the human species started modifying the environment to suit its needs. Although many scientists consider such views oversimplified generalisations, scattered data exist to support them, particularly in the case of plants. Although plants do not have heart beats, plant species that grow fast can be distinguished easily from those that grow slowly. A fast growth rate requires sufficient resources and a corresponding capacity on the plant's part to exploit such resources. As discussed earlier, the resources plants need are extremely simple: sunlight, CO_2 received from the atmosphere, water from the soil, and some

Hasty and patient plants

17 mineral nutrient elements also absorbed from the soil through the plant's root system. The first two resources are stable and guaranteed; their presence throughout the world is approximately the same. The sun rises every day, and the concentration of CO_2 in the atmosphere does not fluctuate widely. However, the same is not true for water and minerals. Therefore, in an environment with insufficient water and nutrient-poor soil, one would expect slow-growing plants to prevail. In such environments, survival is more important than growth, which inevitably occurs at a slow pace. However, when a habitat provides sufficient water and nutrients, the prevailing species are those that grow quickly.

When the environment is unfavourable, survival is more important than growth...

To further this discussion, assume that the properties of fast and slow growth rates are inherent and "written" in the genetic material of respective species. This assumption is only provisional, though, because as seen in Chapter 2, plants – contrary to animals – present much more intense phenotypic plasticity, a phenomenon in which an individual's numerical values for different morphologic, anatomic, physiologic, and biochemical parameters vary significantly depending on the prevailing environmental conditions. Hence, adaptive flexibility is ensured. For now, imagine being in a rich habitat and examine what the plant should do to fully exploit the sufficiency of resources and to translate it into fast growth. Again, to serve the purposes of this discussion, consider only two of the various plant tissues and organs: the leaf, which performs the photosynthetic process, and the xylem vessels in the stem, which transport water and mineral nutrients from the soil, through the roots, and to the leaves. During photosynthesis, CO_2 absorbed from the atmosphere is transformed (reduced) to carbohydrates (sugars). This transformation is not spontaneous but requires energy from the sun (light energy). Chlorophyll, along with other molecules, in an extremely complex intracellular organelle called a chloroplast, absorbs solar energy,

...whereas plentiful resources are conducive to luxurious and wasteful living

which is transformed to chemical energy. The latter, with the help of enzymic mechanisms, helps transform CO_2 into sugars. It is important to understand that for CO_2 to reach the tiny chloroplasts within the leaf cells, it has to follow a specific route. The epidermis of the leaf is impermeable to CO_2, which enters through special entry points formed by special epidermal cells. The gate is internationally called a *stoma* (Greek for *mouth*). In essence, stomata are valves and their opening is regulated by the plant itself, depending on the given potential for effective photosynthesis. Stomata density ranges from a few tens to several hundreds per square millimeter of leaf surface area. In other words, a single plane tree leaf may have 1 million to 2 million stomata. However, because these stomata are tiny, the total percentage of stomata area does not exceed 2 % of the total leaf surface area.

An average leaf has many thousand stomata

After CO_2 passes from the exterior atmosphere to the leaf's interior through the stomata, it enters a maze of corridors, the so-called *intercellular spaces*. In other words, no matter how solid the leaf may appear on the outside, the interior is made up of a collection of cells and free air spaces; parts of the surface of these cells are in contact with one another, whereas the rest of the cells are exposed to the interior gaseous atmosphere. Depending on the plant species, the solid (i.e., cellular) part may make up 60 % to 98 % of the leaf, whereas the rest is composed of gaseous interior corridors so that CO_2 may reach the tiniest, furthest nooks of the leaf.

Leaves are not solid

The more open the stomata and the wider the intercellular corridors, the easier it is for CO_2 to reach the individual leaf cells. Narrow corridors and sparse and half-closed stomata make it more difficult for CO_2 to enter and move in the leaf interior, much as a half-closed door leads to crowding and a very narrow tube leads to resistance against any flow. Every cell receives its share of CO_2, the concentration of which is reduced locally so that the CO_2 removed is replaced through diffusion from the vicinity and, ultimately, from the atmosphere. Fast replacement (i.e., the rate of diffusion) is directly proportional to the photosynthesis rate, that is, the production of sugars, which in turn provides growth with energy and building blocks. For example, sugars produced are transformed into proteins with the addition of nitrogen. Furthermore, through the appropriate biochemical transformations (with the catalytic action of enzymes), sugars may produce lipids, nucleic acids (DNA, RNA), vitamins, polysaccharides, chlorophylls, hormones, and substances related to plant defences (e.g., phenols, terpenoids, alkaloids, steroids). Collectively, these events are called "metabolism" (because one substance *metabolises*, i.e., changes into another).

Therefore, when the habitat is rich, fast-growing plants are expected to prevail, with a leaf structure that maximises photosynthesis. There are numerous stomata with big openings, a light interior structure, and rich enzymic equipment to allow for the reactions of photosynthetic CO_2 assimilation and transformation into a multitude of molecules that participate, functionally or structurally, in plant growth.

Environmental favour or disfavour is reflected in leaf structure and biochemistry...

However, the stomata provide not only an entry point for CO_2 but also an exit point for water in the form of vapour. This water loss in the form of vapour from the aboveground parts of the plant is called *transpiration*. The vapours use the same paths for their course as CO_2, but in the opposite direction. Therefore, wide interior corridors and many large stomata mean not only fast photosynthesis but also fast water loss. If the plant can spare water, this is no problem. Water, as was already discussed, is transported to the leaves through xylem vessels and sucked up by the leaves, as if there were a pump drawing the water upward. As will be seen later, in essence there are two forces contributing to water transport upward, that is, against the direction of gravity. To overcome water's tendency to move downwards because of its weight, a force exceeding that of gravity and in the opposite direction must be applied on the columns of water within the vessels. Without an inner pump similar to the heart of animals, plants manage – by exploiting the physical properties of water – to "draw" it upwards during the day and to "push" it, also upwards, during the night. Although this issue is discussed further later, it suffices to say here that drawing water from a well requires either a suction pump at the surface or a lift pump at the bottom of the well. In both cases, energy has to be consumed. It is noteworthy that for the nightly push of water, plants use a "chemical" pump, or in other words, chemical energy, whereas the daily "drawing" up of water requires no energy consumption by plants, as they simply exploit water's physical properties to make it rise.

. . . as well as the diameter of its vessels

With regard to the fast-growing plant with its stomata open wide and its intense photosynthesis, one must assume that the water supply coming up through the xylem vessels is enough to successfully replace the water that escapes as vapour from the stomata. Consequently, the resistance of the vessels against water transport should be low, that is, their diameter should be large. For example, to draw the same quantity of water into two medical syringes, one with a wide and the other with a narrow needle, more effort must be exerted on the narrow-needle syringe. Alternatively, if the same amount of force is applied in each case, it will take longer to draw the same quantity of water into the latter syringe. According to the mathematical relationship that allows one to calculate the fluid supply of a tube, when the same drawing force is applied at the tube's end, the flow rate increases 16 times when the diameter of the tube is doubled.

The transpiration water current ascending through the xylem vessels also carries and distributes to the plant the necessary soil mineral nutrients, which are absorbed by the roots. Plants need inorganic nutrients because, besides carbon from the atmospheric CO_2 and hydrogen (H) and oxygen (O) from water, the organic compounds produced through sugar metabolism contain various inorganic elements, as the case may be. It is known that proteins also include nitrogen (N) and sulphur (S), elements that usually are transported in their oxidized form as nitrate or sulphate anions, which are reduced in the leaf cells and enter the appropriate precursor substances so that amino acids may be synthesised to merge further into proteins. Proteins are the key to metabolism in all organisms; they organise and regulate all processes within the cell. In plants, the protein concentration in the leaves is positively correlated with photosynthesis and growth rates. In light of the preceding discussion, it is obvious why soil rich in nutrients favours fast growth.

To summarise, a fast-growing plant thriving in a rich habitat needs, among other things, wide vessels, leaves with many large stomata, a loose interior leaf architecture with ample gas spaces (void of cells), and high concentrations of enzymic proteins.

Hasty and wasteful plants have no luck in an alpine environment; they cannot make it when the going gets tough

Let us return, however, to the topic we began discussing a few pages ago: the correlation of growth rate and longevity and the age-old pine trees in the mountains of Colorado. Could a fast-growing plant thrive there? With regard to the necessary resources – light and CO_2 – these are as abundant there as in any other environment. However, the temperatures are low, the growth period is short, and most importantly, the annual rainfall is hardly 250 mm (less than half the mean value for Greece), and the soil is particularly poor in nutrients. Therefore, it is quite unlikely for a fast-growing plant to establish itself there. Its inherent structure would discourage it: its soft leaves would be destroyed by the wind and the collision with tiny windborne ice crystals. The wasteful use of water such a species enjoys in other habitats would leave it dehydrated in an alpine environment. Such a plant could not find sufficient nutrients for its protein synthesis. Its inherent capacity for a high photosynthetic rate would be useless at such low temperatures. Its wide vessels would be susceptible to embolism, that is, the creation of bubbles, which obstruct water flow, inside the vessels. Such embolisms would intensify because of the lack of water and low temperatures, and they would form more easily in the wide vessels of the plant. In this cold, dry environment, the "urge" for fast growth gives way to the need for survival. Plant tissue and organ structure should be robust, to tolerate local conditions. The plant's supporting tissues (xylem, thick cell walls, thick epidermal covering with hydrophobic layers of wax to discourage water loss) account for most of its mass and are the most important investment for long-term survival under such conditions, which would prove too harsh for most organisms. The plant's solid organs seem unconquerable compared with the structurally makeshift construction of dainty, fast-growing plants. Stomata are fewer and smaller, perhaps restricting photosynthesis but also limiting the risk of dehydration. The same is true for the intercellular spaces, which might be small, but fortify the leaf mechanically. The plant's vessels have a small diameter that is suitable for a low water supply, also preventing embolisms and reinforcing xylem strength. In other words, in the plant world (and perhaps in the animal world), leading an intense life is not compatible with longevity.

Ultimately, leading an intense life is not compatible with longevity

As a side note: the previous phrase provides the opportunity to clarify any misunderstanding that may have been created. Readers may have gotten the impression that plants, being inherently fast or slow growing, need nothing but to be in the right environment to thrive. Although basically this is true, one may be misled into thinking that only fast-growing plants are found in a "favourable" environment and only slow-growing plants in an "unfavourable" one. Fortunately, nature is not so predictable. However, before a comment is made regarding the complexity of the concept of "environment" and the epithets *favourable* and *unfavourable*, it may be appropriate to offer a methodologic suggestion: When reading a scientific text, albeit in a popularised form, readers should always compare the writing with their own experience and common sense. Both experience and common sense raise doubts about the statement that only slow-growing plants exist in an "unfavourable" environment, because in almost any environment on Earth, regardless of its level of "favourability," there exist not only many plants but a multitude of various species, big and small, conspicuous and inconspicuous, and slow and fast growing. There is an amazing range of distinct biological properties and specialisations, extending in time and space; in other words, there is *biodiversity*. For example, in a mountainous Mediterranean conifer forest, one's attention may be drawn mostly to the age-old pine or spruce trees. However, if one were curious and adventurous enough to visit this part of the world in early spring, he or she may see crocuses blooming and bursting through the ice, creating a warm oasis where insects find refuge and pay their "rent" by assisting in the plants' reproduction. Later, amidst the mosaic of mud and melting ice, tens of humble-sized species, incapable of forming deep roots, elbow one another to complete their short life cycle as quickly as possible (in a few weeks), before the surface water evaporates into the atmosphere, which is becoming warmer, or recedes into deeper soil layers after the snow melts. At this time, some slightly bigger, annual or biannual plants appear with more competitive roots that reach deeper. These plants also are in a rush to complete their life cycle (i.e., to reproduce and die) before the summer drought. Even biannuals, which can tolerate the lack of water longer, do not stand a chance when the first autumn frost comes and they shed their aboveground parts. Compared with the conifers, all these plants are fast-growing species and succeed one

However, nature is less mundane

Do not let the forest hide the tree

The time sequence of plant forms

another in a continuous procession, alternating colours, smells, and forms, against the permanently dark green background of the respectable slow-growing conifers. They present a variety of biological properties and growth and development plans and strategies so as to fully exploit the potential provided by the resources available in the specific environment.

The time aspect of the environment

What one may conclude from the preceding text is that the environment has a strong temporal component, particularly where the climate presents intense seasonality. This component, combined with plants' plasticity in development and life span (i.e., their capacity to live longer or shorter, depending on the quality of their environment, without losing their reproductive potential), results in the creation of characteristic and unique sets of morphologic, physiologic, and biochemical properties within each species so that the entire habitat is covered spatially and temporally. To return to the previous example, at one extreme there are conspicuous, voluminous, magnificent conifers destined to live for centuries – plants for all seasons – whereas at the other extreme, there are ephemeral, obscure, tiny plants made to exploit part of the environment during a specific season and to end their life cycle within a few weeks – truly straitened species. Between the two extremes, there are many possible intermediate combinations in a century-old, well-tried recipe and in a sequence of events, each stage of which depends on the previous one and leads to the next, a flexible whole that can react and adapt to natural changes while being stable enough to remain undisturbed so that its internal relations do not collapse but evolve. In other words, there is a system that can endure species invasions and departures without risking its cohesion, a system that may well survive the greatest and most sweeping intervention in the history of our planet – which has been in progress for several thousand years and has culminated in the past decades – i.e., that of *Homo sapiens*.

The eternal and ephemeral may co-exist…

…until the overwhelming arrival of Homo sapiens

Plants as Environmental Engineers

Seasonality of climate is not the only thing that allows various organisms to share the same habitat. Another factor is each organism's ability to create its own micro-environment, often a unique set of interwoven physical-chemical conditions that modify space and create an opportunity for a multitude of other organisms to establish themselves and exploit the new situation. In other words, organisms operate as environmental engineers. The best-known example is that of beavers, who turn streams into small ponds by stopping the flow of water. In effect, they add a new dimension to space that is more suitable to their style while their labour is exploited by other organisms that fit the "pond" but not the "stream" system. These organisms, in turn, may cause a series of other modifications. For example, the stream system is less hospitable to several sessile (benthic) algae, which are carried away by running water. It should be noted that unlike higher plant species, algae have no roots by which to anchor themselves to the solid matrix; they simply "stick" to it precariously. This problem is resolved when the water is relatively still. The establishment of benthic, photosynthesising algae enriches the water with organic matter, becoming an attraction for aquatic herbivores; they, in turn, attract their predators and their predators' predators. Then, decomposing organisms arrive on the scene to recycle matter so that more algae may be established in a never-ending sequence of intercorrelation.

Organisms as environmental engineers

In the previous example, the leading actor was the beaver. However, many examples of plants modifying their environment also exist. A group of plants known as phreatophytes can direct their roots to a depth of several tens of meters, reaching the permanent underground aquifer. In deserts or in places with extreme

Bore-drilling plants...

seasonal rainfall, the soil between the surface and the underground aquifer may dry out completely. When the soil is wet (e.g., after the rain), the water fills the tiny interstices between solid soil particles and is subject to counterforces: the force of gravity, which pulls the water down towards the permanent underground aquifer while the capillary phenomenon pushes the water to the surface, where it evaporates into the (permanently thirsty) atmosphere. Although the underground aquifer is a permanent reservoir that can supply some water to higher soil strata, if the system is not replenished by the atmosphere, evaporation and the force of gravity soon prevail over the forces of affinity between water and the soil particles. The soil will dry out and become inhospitable to life. The more the soil dries, the more the resistance against drawing water up through the capillaries increases, because for the phenomenon to actually appear, the gaps first must be "filled" with water. Phreatophytes are the organisms that undertake the task of bringing underground water to the surface. Against the force of gravity, they form a path of least resistance along which water is transported from the underground reservoir to the thirsty surface. Leaves operate as suction pumps, sending the drawn water into the atmosphere through the stomata. This is how, in the middle of the desert, a microenvironment is restored within and around these plants, within which there is sufficient humidity to support life. In the humid undersurfaces of the leaves, fungi and microbe spores will germinate, whereas small animals will find a shady refuge under the plants' canopy.

. . .which create a hospitable environment in deserts so that other organisms may establish themselves

It is believed that the water-tightness of the root may be necessary for water to be pumped up efficiently, yet things are surprisingly less strict just below the soil surface. Therefore, part of the water drawn up is offered by the phreatophyte to refresh the soil around it, thus making it hospitable for the establishment of other plant species. These species usually are not capable of developing roots all the way down to the underground aquifer and depend on the phreatophytes for their survival; the latter function as hospitable hosts, inviting other nearby plants, as if pursuing cohabitation and using their surplus water to pay for it. In the small plant community created, it is not only the *bona fide* guests that benefit. The price paid by the colonists is that they offer a kind of passive protection to their host. If the founder of the plant community were alone, it would be much more at risk of attack by its consumers. The presence of guests spreads the risk, because there are more prey from which the predator may choose. Ultimately, the established phreatophyte is for its microcosm what an oasis is for Bedouins in the desert. The example of the beaver is more easily understood because it presupposes a series of voluntary and programmed activities on the beaver's part, similar to those humans engage in to modify their environment with an ultimate purpose in mind. In the case of desert phreatophytes, however, the action is involuntary. Plants change their environment simply because they happen to be there. Indisputably, despite the lack of intent, the results are highly significant at both the microcosmic and the global level, as will be shown in the next pages.

Involuntary modifications of the environment by plants...

There are many more examples of environmental modifications with local importance. The trunk of a tree in the northern hemisphere has a southern, well-lit, and warm aspect and a northern, shady, and cooler aspect. The difference in surface temperature between the two sides may be as great as 15 ° C. The tree canopy creates shade and therefore makes it possible for plants that do not tolerate intense light to grow. Under the leaves and bark cracks, many insects and small mammals may find refuge and hide from their enemies. The leaf surface alone, with its pronounced microscopic relief, as well as the hairs and the complex chemistry of the epidermal cover (cuticle), comprises a special ecosystem inhabited by several microorganisms and microscopic invertebrates. Water evaporation through leaf stomata enriches the atmosphere with water vapour, thus increasing relative humidity around the plants. Furthermore, vegetation significantly reduces wind speed.

...and modifications with characteristics of target pursuing: chemical modification of the soil and choosing neighbours

In the examples presented, plants behave as passive modifiers of the natural environment. In other cases, however, chemical substances secreted by plant roots into the soil or secreted by the leaves and then precipitating into the soil through rain may be toxic to certain plant species, thus preventing their seeds from germinating and new, antagonistic organisms from being established. The same substances may favour the germination of other plant species seeds, thus helping their establishment. In this way, plants shape a chemical environment that blocks the establishment of certain neighbours while favouring the establishment of others. For the former, this is a chemical war, for the latter a chemical facilitation. To a certain extent, then, plants choose their neighbours. In the same chemical manner, plants attract from the huge number of soil microorganisms those with which they can establish a symbiotic relationship, that is, companies for mutual benefit, in which each member provides what it can spare to its partner and receives what it needs. Symbiosis between plants and microorganisms is the rule in nature, but it escapes our attention because it usually occurs underground. Such symbiotic relationships are of the utmost biological and ecological interest; they contribute to nitrogen recycling and the underground transport of chemical elements and compounds through an extensive and complex network in which the role of terminal stations is undertaken by plants and the role of wiring by the fungal hyphae. This concept will be examined further in Chapter 8.

Chemical invitation to establish underground companies with microorganisms

Quite often, the substances released are volatile and odorous; therefore, they can arouse animals' olfactory mechanisms, transferring a suitable type of information – for example, that there is a flower nearby and therefore food in the form of pollen or nectar. This is another issue that will be discussed later, in Chapter 5. For now, we may conclude that plants involuntarily modify their physical environment and that this modification is exploited by other organisms. At the same time, plants modify their environment chemically and this modification seems to have the characteristics of a specific "goal." In the latter case, an additional materials and energy cost is required for the biosynthesis of the particular substances that will intermediate to ensure the aforementioned goals are achieved.

Local modification of atmospheric chemistry and invitation to pollinators

In all these cases, the environmental modification is of a local nature; it concerns the plant's immediate vicinity. However, there also is a plant function with a global impact. As will be shown later, plants play the most significant role in creating and maintaining the chemical composition of the earth's atmosphere.

Environmental modifications of a planetary scale

Currently, the atmosphere is 79 % nitrogen (N_2) and 21 % oxygen (O_2). There also are traces of two other, very significant gases, carbon dioxide (CO_2) and methane (CH_4), as well as water vapour. Life on earth used available chemical elements and compounds as raw materials; it combined them and transformed them to serve its purposes. This happened not only with the elements provided by the planet's crust, or those dissolved in water, but also with atmospheric gases. As a result of this (bio)chemical process, variations occurred in the concentrations of specific atmospheric gases, closely related to plant evolution and functions. In other words, the atmosphere was not always as we know it today. Its history was shaped in parallel with that of plants.

The earth's atmosphere was not always as we know it today...

It is assumed that the reader is familiar with the prevalent biological view that present-day organisms are products of the evolution of older organisms through the process of natural selection. Very briefly, this means that small, random changes in genetic material (i.e., mutations), which sometimes occur very slowly and sometimes are accelerated, consequently lead to small changes in the form, behaviour, or properties of an organism. Such changes may render the organism stronger (i.e., better adapted) or weaker compared with other individuals of the same species living in the same environment. In addition, there is a third case, in which such changes are neutral and offer the mutant neither an advantage nor a disadvantage. Hence, the mutation is subject to examination; it is judged favourable or unfavourable for the specific environment. Favourable changes are maintained because their carriers leave many more offspring; therefore, the mutated gene is dispersed further and increases its frequency in the population. The unfavourable change is eradicated because its victim leaves behind fewer offspring. A neutral change remains latent and may prove favourable or unfavourable later, when the environment changes (i.e., when there is a change in the criteria that will make it one or the other).

...nor were organisms the same

Mutations and evolution

Mutations are subject to the discretion of natural selection, and favourable ones produce more descendants

The "target" of evolution, therefore, is the individual within the population of a species. A biological species is defined as the set of individuals that can reproduce with each other but not with individuals of another species. Therefore, genetic exchange normally occurs within a species. Consequently, a "species" is a mosaic of individuals that have similar basic traits, allowing them to intermingle, and have a sufficient variation of properties (gene total) so that natural selection (the environment) is allowed to impose criteria for success. Many times, a specific "set" of properties is established within a population in the form of a breed or variety. At a consequent step in the evolutionary process, reproductive obstacles might occur that may isolate such

A definition of species

Reproductive impediments and the birth of a species

a variety from the rest of the population. The most common reproductive obstacle is geographic isolation. For example, the founders of a colony on an island may have a genetic composition slightly different from that of the average population, or they may comprise a separate variety. Their reproductive isolation from the rest of the population due to distance would result in a deviation in genetic frequency, which would be affected further by new, random mutations, different from those of the rest of the population. Because there is no possibility of mixing the new forms and properties with the older ones, the deviation in genetic frequency leads to the creation of a new species. The accumulation of new properties (related not only to the organism's form but also to its physiology and chemistry) ultimately makes it impossible for new individuals to reproduce with their ancestors, even if the geographic isolation were removed. Thus, a new species is born. This process is long, depends on the type of organism, and seems to be accelerated when environmental changes also occur quickly. In any case, it occurs over hundreds of thousands to millions of years.

The role of geographic isolation and the diverging genetic frequencies

These issues will be examined further, when the evolution of plants is described in Chapter 4. However, it is important to be aware that the matter of evolution is not a theoretic intellectual construct. Although nobody was alive to describe it, fossils – those wonderful time machines – indicate that the world was not always as it is today. This is true not only of organisms but also their environment, including the chemical composition of the atmosphere. This composition has left its traces on the beds of rock on the earth's crust. Furthermore, rock beds and fossils can be dated, so geochemists know that certain ancient rock beds could not have been formed if the earth's atmosphere had the same composition it has today. Astrophysicists make reasonable hypotheses about the earth's age and the first stages after its formation. Humans now are in the position to

Fossils and rocks: machines running through time

record with logical consistency the wonderful history of life on our planet. Indisputably, there are gaps. Yet, this history does not contradict the knowledge we have accumulated about matter, nature, and organisms during the past 300 years of scientific revolution, which is still occurring and will continue to the extent that the prevailing species on Earth proves intelligent and prudent enough to preserve the world it already started destroying.

If plants stopped photosynthesising, atmospheric carbon dioxide would double in 10 years and oxygen would almost disappear in 4,500 years

Atmospheric CO_2 and O_2 are continuously recycled through organisms, the physiologic functions of which absorb and release the two gases. It is estimated that all the atmospheric CO_2 (the concentration of which is 0.037 % today) is recycled by going through living organisms over a period of 10 years. O_2 (the atmospheric concentration of which is 21 %) similarly is recycled within 4,500 years. If there were no organisms, the corresponding inorganic recycling would take at least 700 times longer. The first conclusion, therefore, is that the composition of the atmosphere is unstable but its balance is maintained by the presence of living organisms.

Gas exchange between organisms and the atmosphere keeps the latter relatively stable

The functions in which CO_2 and O_2 participate, either as reactants (whereupon they are removed from the atmosphere) or as products of a reaction (whereupon they are released into the atmosphere), are photosynthesis, respiration, and photorespiration. Plants, algae (whether benthic or drifting planktonic), and certain aerobic photosynthesising bacteria perform all three functions. Animals and certain aerobic non-photosynthesising bacteria perform only the second function, respiration.

A brief and hardly technical description of photosynthesis

During photosynthesis, atmospheric CO_2 is absorbed by plants and, with the help of the energy from light, reduced into sugars. The reaction starts with the absorption of light by chlorophyll. Through a complex process, the energy from the light absorbed is successively transformed into electric and, finally, chemical energy in the form of a potent phosphate bond within a molecule called adenosine triphosphate (ATP). In return, ATP helps a series of enzymes to act as catalysts in reducing CO_2 into sugars. The energy contained in a sugar is much greater than that of the corresponding CO_2 molecules from which it originated. In other words, during photosynthesis, the energy of light trapped by the chlorophyll is stored in the sugars created from CO_2.

An oversimplified presentation of the reaction is as follows:

$$CO_2\downarrow + H_2O \xrightarrow{\text{Light, chlorophyll}} \text{sugars} + O_2\uparrow \quad (1)$$

Ultimately, all the food on Earth is produced by plants...

It is clear that water participates in the reaction and O_2 is produced as a by-product. The downward-pointing arrow means that CO_2 is fixed and removed from the atmosphere, whereas the upward-pointing arrow indicates that O_2 is released into the atmosphere. This reaction takes place within special leaf cell organelles called *chloroplasts*. One may imagine these organelles as factories that import energy (light) and simple, unprocessed raw materials (CO_2 and water) and produce a useful, high-value product (sugar) to be exported. Inevitably, however, toxic waste also is produced and must be removed. Indeed, O_2 is a by-product of the process and should not accumulate in large quantities within the chloroplast because it becomes toxic in high concentrations. Sugars are exported to the rest of the plant and have a dual role to play: On one hand, they are the construction blocks for the synthesis of thousands of other substances required by the plant. With suitable additives and transformations, with the sugars produced through photosynthesis used as an initial resource, other substances are formed, such as proteins, lipids, other sugars, polysaccharides, cellulose, lignin, wood, vitamins, plant hormones, alkaloids,

...as is atmospheric oxygen

Retrieving the energy content of food is achieved through respiration, which is performed by both animals and plants...

chlorophylls, carotenoids, and phenols – in other words, the entire plant body. On the other hand, sugars also may be broken down into their components (i.e., CO_2 and H_2O), thus releasing the energy they contain (which they acquired through photosynthesis) wherever it is needed, such as in plant tissues incapable of photosynthesis. This sugar breakdown requires the cooperation of two cellular compartments: the so-called cytoplasm and the mitochondria. The process is gradual and strictly controlled; it is called *respiration* and is represented by the following equation:

$$\text{Sugars} + O_2\downarrow \longrightarrow CO_2\uparrow + H_2O \qquad (2)$$

...and during which oxygen is consumed

From a technical point of view, respiration follows a process more or less similar to photosynthesis, without the involvement of light. Therefore, when glucose breaks down, it releases energy, the main part of which is transformed successively into electrical and, ultimately, chemical energy in the form of ATP. The carbon contained in the sugars is discharged into the atmosphere in the form of CO_2, but O_2 is required for the reaction to take place. This is the same O_2 animals need to breathe. Respiration, after all, in its fundamental, cellular aspect, is the same in all aerobic organisms.

In practice, O_2 is necessary as the final recipient of the electrons produced during the process of cellular respiration. O_2 receives the electrons and protons (hydrogen cations) and is reduced to water. Reaction (2) is similar to – and often is mistakenly called – the burning of sugars. The difference is that sugar "burning" (like the burning of any other organic substance) does require O_2, but it is a violent and rapid process. In a sense, cellular respiration indeed "burns" sugars but does so in an organised and controlled manner so that the energy contained is not all transformed into heat, as in actual burning; its greater part is retrieved as useful energy in the form of ATP, which is to be used to cover the energy needs of the cell.

Photosynthesis and respiration are reverse reactions: if they are performed at the same pace, the composition of the atmosphere remains stable. This is not always the case, however.

However, looking at things from our initial goal, that is, the involvement of organisms in the exchange of gases with the atmosphere, we may observe that the reactions of photosynthesis (1) and respiration (2) are the exact reverse of each other with regard to absorbing from and releasing to the atmosphere the gases involved. It also should be reiterated that while plants perform both processes, animals perform only respiration. The organic substrate for the respiration of herbivorous animals is taken with their food from plants. The equivalent for carnivores is taken from the herbivores that have consumed plants. Ultimately, all food comes from plants.

Additionally, plants perform a third – quite strange at first glance – exchange of gases known as *photorespiration*. The second half of the word implies a process in which the final result is the same as that of respiration: the absorption of O_2 from the atmosphere and the release of CO_2. However, the biochemical processes, as well as the enzymes and organelles involved in photorespiration, are completely different. The first half of the term implies that it is performed only in the presence of light, whereas regular respiration takes place in both light and darkness. Quantitatively, photorespiration contributes significantly to the exchange of O_2 and CO_2 between the plant and the atmosphere. Its ultimate significance, however, seems to go far beyond a mere quantitative contribution. As will be observed later, photorespiration is the valve through which plants ensure that atmospheric concentrations of CO_2 and O_2 remain within appropriate limits for their growth and survival.

If something goes wrong, plants intervene with a biochemical safety valve to restore the atmospheric composition that suits them

Chemical History of the Atmosphere: Photosynthesis and Plants as the Main Players

The history of the earth is divided into geologic aeons. The transition from one to the next was accompanied by fast changes in the natural environment and organisms...

The earth is believed to be 4.6 billion years old. Its age has been divided into the so-called geologic aeons; aeons are divided into eras and eras into periods. This is not an arbitrary division. Every subunit is characterised by a specific phase in the evolution of the natural environment and of organisms. Furthermore, with regard to the last three aeons – the Palaeozoic, Mesozoic, and Caenozoic – each change from one to the next was marked by dramatic episodes. The Palaeozoic aeon started (546 million years ago) with the fast and abrupt appearance of multicellular animals, whereas the transition from Palaeozoic to Mesozoic aeons (248 million years ago) and from Mesozoic to Caenozoic aeons (65 million years ago) was marked by massive extinction of animal species. Although this chronology will be followed for the purposes of our analysis, it must be said that it has been constructed totally on the basis of animal rather than plant criteria. As the next chapter explains, mass extinctions of animal species were not accompanied by mass extinctions of plant species, at least not to the same extent. However, findings of fossilised animals are much more numerous, because the presence of a hard internal or external skeleton increases the likelihood of fossilisation. Besides, fossilised animals are indisputably more impressive than fossilised plants. The aforementioned chronology, which is based on animal fossils, is the oldest and best known; this is the reason for using it. Moreover, a chronology based on plant criteria would be quite different.

... usually mass extinction of animals...

...but not of plants

The earth's crust was solidified and the first oceans appeared 4 billion years ago

During Earth's early years, its environment was particularly inhospitable. It is believed that the earth's crust solidified from the original fiery mass around 400 million to 600 million years after the creation of the planet and the gradual drop of its temperature. It was around that time that the first oceans were created from the condensation of water vapour in the atmosphere, and possibly also from water from colliding comets. This is particularly significant because life is closely related to the presence of water; at that time, though, the temperature of the oceans was still too high at 80 °C to 90 °C. The atmosphere contained little O_2, far below 0.001 %, as compared with 21 % at present. As for CO_2, the concentration was 5 %, or at least 100 times that of today, which caused a huge-scale greenhouse effect, although the sun might have been emitting 25 % less energy at that time. Volcanic activity, facilitated by the particularly thin young crust of the earth, kept enriching the atmosphere with CO_2, methane, and other greenhouse gases. Furthermore, there were numerous meteorites within the still unstable solar system; these kept colliding with the young earth, intensifying the phenomenon and causing repeated evaporation of the oceans due to the heat of collisions. The moon had just separated from the earth. Meteoric collisions with the moon during that time left their marks; the resulting craters are still visible on its surface. These craters were not eradicated because the moon, with its small mass, could not maintain an atmosphere; therefore, detrition could not occur.

However, the temperature was very high and there was no oxygen in the atmosphere

For another 0.2 billion years, the earth suffered vehement meteorite attacks

On the contrary, because the earth managed to maintain an atmosphere due to its higher gravitational force, the marks from the early meteorite attacks were soon worn away. The calmness that followed allowed the earth to cool gradually to temperatures that made the appearance of life possible. If for a moment that huge question of how life appeared on our planet (i.e., the question of origin) is ignored, it may be said that life appeared quite suddenly 3.8 billion to 3.6 billion years ago in the ocean. One may be led to this conclusion by the changes in the composition of rocks created at that time, as well as by the first fossils of tiny bacterial cells dating back to the same time. Life appeared in the oceans and remained there for about 3.2 billion years,

Calmness started around 3.8 billion years ago and, surprisingly, the first microbial organisms appeared almost immediately

Life appeared in the oceans, and 3.2 billion years had to pass before the land was colonised

but not because there was no land. The first continent on the planet may have appeared before life on Earth. Imagine the earth as a huge rocky mass without soil, because soil is created mainly as the result of the effect of plants. Life was not possible out of the water, mainly because the stratospheric ozone layer, which absorbs harmful ultraviolet radiation, did not yet exist. This radiation is absorbed by critical biological macromolecules, such as DNA and proteins, and causes fatal damage. The ozone that absorbs such radiation resulted from the reaction of this radiation with O_2, yet O_2 was practically absent from the early atmosphere. The explanation that follows indicates that the activity of photosynthesis led to an increase in oxygen; this gradually created – on the "ceiling" above the earth's atmosphere – the protective ozone layer. In addition, photosynthesis caused a decrease in CO_2 and this, in turn, mitigated greenhouse effects and allowed milder temperatures to prevail.

Why was land inhospitable?

In the absence of O_2, the metabolism of the first organisms had to be anaerobic. As with present-day anaerobic microorganisms, O_2 also had a harmful effect on the first ones. Their biochemistry and physiology were adapted to survive the lack of O_2, and there was no reason for these organisms to develop protective mechanisms against the toxic effects of O_2 (all present-day aerobic organisms have such mechanisms). Although today O_2 is absolutely essential for respiration, it is potentially dangerous, either by itself or in the form of certain of its metabolic products, which are called toxic radicals. Such radicals oxidize and destroy biological membranes and have a negative effect on DNA if the defensive mechanisms that neutralise them underfunction. On the other hand, aerobic metabolism is much more effective from an energy point of view than anaerobic metabolism. Therefore, in a sense, aerobic organisms are on the razor's edge: they cannot live without oxygen, but they need to limit its toxic effects.

The first organisms were anaerobic

Oxygen is useful for respiration, but it becomes toxic if its oxidising action is not restricted

It is a bit like nuclear energy, which might be useful but also might get out of control. There must have been many natural experiments before the perfect system emerged that could both exploit the efficacy of oxygen (as the ultimate acceptor of electrons during respiration) and harness the uncontrollable, destructive, chain reaction of its toxic radicals. Some experts think that several degenerative ailments of organisms, including ageing, may be related to the gradual and possibly inevitable accumulation of damage caused by toxic oxygen radicals because of the collapse of corresponding defence mechanisms. Consider the popularity of dietary antioxidants among humans; even if there is some marketing exaggeration involved, a significant body of evidence exists indicating that damage to an organism's defence against oxidation might be harmful.

The first photosynthetic bacteria did not produce oxygen

With regard to the anaerobic environment, in which the first microorganisms appeared, although some of them may have been capable of photosynthesis, they must have adopted a type of photosynthesis that did not produce O_2 as a by-product. Even today, there are photosynthesising anaerobic bacteria; in the basic photosynthesis equation (1), H_2O is replaced by other compounds (e.g., hydrogen sulphide [H_2S]), whereupon instead of O_2, sulphur (S) is released as a by-product into the environment.

$$H_2S + CO_2\downarrow \xrightarrow[\text{Bacterial chlorophyll}]{\text{light}} \text{sugars} + S \qquad (3)$$

Then came photosynthesising cyanobacteria, which produce oxygen

Later, cyanobacteria appeared; these are photosynthesising microorganisms that produce oxygen during their photosynthesis, just like algae and plants. Today, they float like plankton in all water masses, whether marine or lacustrine. Their earliest fossils date back 2.7 billion years, which is the latest time point at which they may have appeared; they probably appeared earlier, but there are no cyanobacterial fossils from that time. This view also is corroborated by the gradual appearance of oxidised rock beds dating back to the same period, indicating a timid and gradual appearance of atmospheric oxygen. It is believed that the basic threshold of an O_2 concentration equal to 0.1 % was exceeded 2.2 billion years ago, whereas 2 billion years ago the stratospheric ozone layer gradually made its appearance.

Anaerobic photosynthesis is of local significance. It can take place only where H_2S exists.

It is worth noting some characteristic differences between equation (3) (anaerobic photosynthesis) and equation (1) (aerobic photosynthesis). Equation (3) represents the activity of anaerobic photosynthetic bacteria, whereas equation (1) reflects the activity of cyanobacteria, algae, and plants; (3) is older than (1) and requires that H_2S, which is relatively scarce, be present. On the other hand, (1) requires water, which is much more available, in place of H_2S. Even if H_2S was much more available in the first anaerobic stages of life evolution, unlike water it was not found everywhere. Therefore, anaerobic photosynthesising microorganisms had to live near points where H_2S was released, such as cracks in the earth's crust and warm springs gushing sulphureous water. If they distanced themselves from such sources, they would starve. By replacing H_2S with H_2O, photosynthesis could feasibly spread everywhere on Earth, provided there was light and water. Indeed this did occur everywhere. Cyanobacteria dominated the earth and made every tiny drop of water mass productive. The production of organic matter by the photosynthesis of these organisms also facilitated the evolution of heterotrophic microorganisms. Following the appearance of aerobic photosynthesis, life no longer was limited to localities around geologic cracks or warm springs. It was the first time in the earth's history that life was globalised, at least in the oceans. The land environment, however, remained uninhabited. It would take another 1.5 billion years after the dominance of cyanobacteria and a series of

Oxygen-producing photosynthesis can take place everywhere, because it uses water – which is ample – as a substrate

This is how food production was globalised for the first time

important evolutionary steps for plants and other multicellular organisms to appear, when 450 million years ago, plants migrated from the water onto land. The most important of these steps are described in the next chapter.

In the meantime, the concentration of O_2 reached 10 % (about half what it is today), whereas that of CO_2, although reduced to 0.5 %, was still 12 to 13 times higher than it is today, 0.037 %.

The appearance of oxygen in the primal atmosphere was one of the most intense incidents of atmospheric pollution...

What impact did the appearance and gradual increase of O_2 have on the evolution of life? First, for the organisms of that time – which initially were microscopic and single-celled and later were made up of few cells, although still microscopic – the appearance of O_2 must have been an unpleasant environmental and metabolic surprise. A new, intensely oxidising molecule suddenly appeared in a reductive world; a penetrating molecule suddenly started oxidising biological structures that had been created on the basis of its absence. It must have been one of the most extreme episodes of atmospheric pollution in the earth's history. *Mutatis mutandis*, it must have been as if organisms producing cyanide (HCN), a gas known to be poisonous to aerobic organisms, appeared today. Only two solutions would have been possible: either organisms would have had to adapt to the new conditions by developing biochemical mechanisms to protect and detoxify themselves from the poison (i.e., mechanisms to transform the drastic substance into an inert one), or the organisms would have become extinct. When oxygen appeared, both events must have taken place. Besides the organisms that disappeared because of the new apocalypse, there also were more versatile organisms that gradually developed antioxidising systems. The wealth of these systems is seen in present-day organisms, particularly photosynthesising ones. The changes mainly concern metabolic cycles through which – in the presence of appropriate enzymes – the

...which must have left numerous victims in its wake...

...before biochemical antioxidant systems were evolved

toxic forms of O_2 are reduced at the expense of another substance that is oxidised. The substances produced are less toxic. Furthermore, through suitable transformations, the substances produced may be reused to capture and neutralise new toxic radicals. It is now known that substances such as ascorbic acid (vitamin C), tocopherol (vitamin E), and flavonoids are critical chemical substances forming the basis of plant antioxidising systems. These are substances animals borrow by eating plants or by maintaining in their digestive tracts populations of beneficial bacteria that produce these substances.

Indisputably, the first organisms to solve the problem of oxidation were photosynthesising cyanobacteria. Resolving this problem was a prerequisite for the development of aerobic photosynthesis, which produces O_2 within the cell. As will be seen later, in the chapter about plant evolution, the capacity of photosynthesis and of protection against the oxygen produced was passed down by cyanobacteria to other non-photosynthesising cells so that organisms similar to algae emerged (endosymbiotic hypothesis). Later, plants originated from the algae.

An intelligent response or how a dangerous pollutant becomes exploitable

Evolution, though, proved to be truly intelligent. It did not limit itself to neutralising O_2 but also created mechanisms to exploit it. Before O_2 appeared, the energy efficiency from breaking down an organic or inorganic molecule that anaerobic organisms found in their environment was very low. So-called anaerobic respiration (which is still used by anaerobic bacteria) is based on the partial oxidation of respiratory substrates through the removal of protons or electrons. Its energy efficiency was and still is poor, and the products of oxidation must be excreted from the organism as useless or, sometimes, toxic. Series of bacterial species appeared (and still exist), in which the successor received the excrement of the previous one to further oxidise it into a third product, which in turn was taken

Anaerobic respiration had very poor energy efficiency, and end products ranging from useless to toxic had to be removed

up by the next successor. In that way, every member of this sequence of microorganisms faced the problem of low energy efficiency as well as that of safe waste removal. The solution of recycling through the intervention of organisms, each one eating the waste of the previous one, was quite clever. What was even more clever, though, was the use of the waste, i.e. oxygen, as the final acceptor of protons and electrons produced from the oxidation of organic substances. First, this multiplied energy efficiency and, second, it resolved the issue of waste. The technical explanation for this increase in energy efficiency is beyond the scope of this book; however, it suffices to say that this resolution of the waste problem was truly inventive. The uptake of protons and electrons produced from the oxidation of respiratory substrates by O_2 releases water, a harmless product that hardly needs to be excreted from the cell, as 90 % to 95 % of the cell is made up of water.

Aerobic respiration, with oxygen as the electron acceptor, led to soaring energy efficiency in the oxidation of sugars, and its end product, water, is harmless

The increased energy efficiency of respiration was the deciding factor in the evolution of animal organisms. As early as the beginning of the Cambrian period (550 million years ago), there was an amazing variety of marine animal species in large populations. The basic organisation of the body was already completed in the classes of the invertebrates, whereas the higher evolutionary classes of fish, amphibians, reptiles, birds, and mammals appeared – in that order – in the next geologic aeons. Free from energy limitations, animal organisms could now channel their surplus energy into movement and the development of bigger bodies. The need of the first multicellular animal organisms for defensive armour led to the creation of an external skeleton, similar to that of present-day corals, snails, shells, and so on. The basic raw material for this skeleton is calcium carbonate ($CaCO_3$), which is formed from calcium oxides and CO_2, ultimately

The energy surplus of aerobic respiration could be channeled to animal movement and more intense metabolism

Photosynthesis by cyanobacteria and algae and the formation of the external skeleton of the first invertebrates contributed to the alleviation of the ancient greenhouse effect

originating from the atmosphere. In other words, CO_2 was reduced from its initial level of 5 % to 0.5 % at the beginning of the Cambrian period not only because of its absorption into organic matter by photosynthesising organisms, but also by its absorption into inorganic $CaCO_3$ in the external skeleton of primordial multicellular animal organisms. This is the first significant contribution animal organisms made to the evolution of the earth's atmospheric composition. The second has been taking place for the past 300 years: it is called "anthropogenic atmospheric pollution" and only one animal species is responsible for it.

In the past 450 million years, the main player in the regulation of atmospheric composition has been land plants

The subsequent history of the evolution of the earth's atmosphere from 450 million years ago until now is related – if not exclusively, then at least mainly – to the activity of land plants.

Colonisation of the land by plants was not easy. In the next chapter, there is a description of the physical and biological prerequisite conditions and the anatomic and physiologic innovations necessary before land could be conquered. Let us assume this happened 450 million years ago. In the next 50 million years, the barren, lunar landscape of the earth was replaced by a green world, covered by every form of vegetation, including forests. Of course, as soon as plants colonised the land, animals soon followed. The amphibians were the pioneers, and they evolved to adapt to this dual life: partly in and partly out of water.

The chemical history of the atmosphere

After the land turned green, the adjustment of atmospheric concentrations of O_2 and CO_2 was performed mainly by land plants. The conquest of land expanses soon increased the atmospheric concentration of oxygen from 10 % to 12 % to the unusual and dangerous level of 25 % to 30 %. Correspondingly, CO_2 dropped from 0.5 % at the beginning of the Cambrian period to a minimum of 0.05 % during the Carboniferous period of big forests. Whereas the increase in O_2 was the result of photosynthesis, the reduction of CO_2 was partly the result of the soil detrition caused by plants forming roots. Although a similar type of detrition may take place abiotically, its pace is much lower. To develop roots, plants have to penetrate rock beds, which is why they excrete organic acids that make room by eroding the rock. At the same time, the rock bed releases inorganic elements useful to the plant (e.g., phosphorus, iron, magnesium); these are absorbed by the roots. Part of the nonabsorbed elements are carried by rain to the sea, where they act as fertiliser for the growth of algae, which through photosynthesis further decrease the atmospheric level of CO_2. However, the excretion of organic acids from the roots intensifies the transformation of complex calcium/silicon oxides ($CaSiO_3$) into simple silicon oxides. This is how atmospheric CO_2 participates in the equation:

Both photosynthesis and the formation of roots contribute to alleviating the greenhouse effect

$$CO_2 \downarrow + CaSiO_3 \longrightarrow CaCO_3 + SiO_2$$

Therefore, plants contribute towards reducing atmospheric CO_2 not only through photosynthesis, but also through the capacity of their roots to penetrate and erode rock beds and create soil.

High O_2 concentrations during the Carboniferous period were hazardous because they caused spontaneous fires. Indeed, there are many indications (among them the coal ores created during this period) that huge fires destroyed a significant part of the vegetation. The combination of combustion, which consumes O_2, with the reduction in vegetation, which produces it, led to the correction in O_2 concentration to lower levels, stabilised at 20 % to 22 % during the past 170 million years. As for CO_2, its values ranged between 0.25 % and 0.028 % after the Carboniferous period (i.e., in the past 250 million years). These fluctuations were determined, to a large extent, by photosynthesis and the state of vegetation; however, they also have been affected by violent natural phenomena, such as volcanic eruptions and the

Why should the oxygen concentration not exceed 25 %?

fall of large celestial bodies. For example, at the end of the Permian period (250 million years ago), the tectonic movements of the earth's crust created a supercontinent called Pangea. The huge volcanic activity accompanying it is believed to have multiplied the concentration of CO_2 six times within about 10,000 years. The fall of the meteorite at the end of the Cretaceous period (65 million years ago), which meant the end of the dinosaurs, increased CO_2 by 20 %. In the past 400,000 years, CO_2 has decreased to its lowest figures in the earth's history, ranging between 0.018 % and 0.028 %. The manmade CO_2 increase following the Industrial Revolution (by about 50 % in the last 300 years) occurred several times faster than that caused by volcanic activity during Pangea's creation. Then, climatic changes led to the extinction of 90 % of animal species (but not plant species; the reasons for this difference are explained later). The question remains as to what the outcome of imminent manmade climate changes will be.

Volcanic action, meteorites, and climatic changes of the past

Present-day manmade climatic changes are faster

In conclusion, the photosynthetic activity of (mainly) land plants and the soil detrition caused by their roots not only led to present-day concentrations of O_2 and CO_2 in the atmosphere, but also to the counterbalancing of these concentrations and their maintenance within a range acceptable for life, when natural or biological factors tend to change it. With regard to the topics discussed previously, one may reasonably conclude that the modification of atmospheric composition by plants is a gigantic phenomenon of a planetary scale with a huge impact on the life of all organisms. Remember that small-scale modifications in the immediate plant environment have been classified as involuntary or passive (i.e., modifications that take place simply because plants exist) and "voluntary" (so to speak) modifications at some energy cost for plants in

Not only did plants contribute towards modifying the earth's atmosphere...

pursuit of a "goal." Although this might appear to be an exaggeration at first sight, the planetary-scale counterbalancing of atmospheric composition does not seem to be a very random phenomenon. It is associated with the appearance and maintenance of specific biochemical characteristics of photosynthesis, which, at the expense of plant productivity, keep the atmospheric gases involved (O_2, CO_2) within limits that are compatible with the sound health of the planet.

...but they undertook its maintenance on a global scale, in a paradoxic contract work scheme

Planetary Aspects of an Enzymic Reaction

The assimilation and reduction of CO_2 into sugars is a complex biochemical process catalysed by enzymes and fuelled by chemical energy originating from photosynthesis. The initiating enzyme is called ribulose 1,5-bisphosphate-carboxylase/oxygenase. This enzyme holds two world records. First, all the carbon of the biosphere – in all the bodies of all organisms, without exception, contemporary and past ones, whatever their tier on the tree of life – has passed through the reaction centre of this enzyme, that is, it has been processed by this specific enzyme. Second, it is by far the most multitudinous enzyme on the planet. Because it is so famous and important, it has been given the nickname *Rubisco*. Rubisco captures carbon dioxide (which has one carbon atom) and adds it to a receptor molecule, namely ribulose 1,5-bisphosphate, a sugar with five carbon atoms. (There are sugars with three to a few thousand carbon atoms. Ordinary table sugar has 12 carbon atoms in its molecule; honey glucose and fructose each have six, whereas starch has a few thousand.) This action results in a compound with six carbon atoms, which is unstable and therefore immediately breaks down into two molecules of an acid (phosphoglyceric acid) with three carbon atoms each.

Rubisco: an enzyme of global importance

About enzymic metabolic cycles...

Other enzymes then intervene, receive the phosphoglyceric acid, and through a series of transformations fuelled by energy produced via photosynthesis, transform it into a series of useful sugars on one hand and the molecule receptor of the carbon dioxide – ribulose 1,5-bisphosphate – on the other. When repeated six times, this cycle produces one molecule of glucose from six molecules of carbon dioxide without actually consuming the carbon dioxide receptor.

...which operate like factories, importing simple, unprocessed raw materials and energy to export complex, processed products...

Imagine the enzyme system of CO_2 assimilation as a factory that imports a raw material (CO_2) and processes it with two objectives: first, to export sugars and, second, to recompose the receptor. The enzymes and intermediate paths channeling the carbon dioxide are designed to ensure smooth and regulated circulation, depending on the availability of the raw material (CO_2) from the atmosphere and energy from photosynthesis. In other words, the system is organised so it may tolerate a temporary shortage of raw material (e.g., when leaf stomata are closed and the carbon dioxide cannot enter the leaf) or in cases of an energy crisis (for example, when the sky becomes dark). The factory enzymes – the engines recognising the structure of the molecules involved, coordinating the course of events, and regulating the pace of change – receive signals (information) regarding the levels of available light and the potential of its efficient use by photosynthetic membranes for the production of chemical energy. This is how enzymes regulate the overall pace of product production and modify the circulation of substances in the maze-like interior network of the factory, depending on the energy and raw materials available and on the demand for the end product.

...and are regulated by supply-and-demand laws

Rubisco is a strange enzyme, however. To fully comprehend this statement, one needs to consider some basic principles of enzyme function. Enzymes are considered specific, that is, they catalyze only one reaction or a very small number of reactions.

This is necessary for metabolism to have an organised form. In other words, enzyme A can catalyse the reaction of transforming substances a and b into substance c (or vice versa) and only that. For substances a and b to react (and produce product c), they must come into contact. Enzyme A provides the venue for this encounter. This venue is called the enzyme's reaction centre. To keep irrelevant substances from interfering and causing confusion, this centre is built to receive only the substances that will react, and only these. The relationship between an enzyme's reaction centre and the substrates of a reaction often is compared with the relationship between a key and its keyhole. Enzymes are huge molecules (compared with their substrates) with a special spatial structure, within which there is a place where the molecules-to-react – that is, a and b – can approach and bond. In practice, of course, substances with structures similar to those of a and b may deceive the reaction centre, as a master key would fit a keyhole, and prevent – more or less successfully – the approach of a and b, thus inhibiting the reaction. This is why the world *affinity* is used to describe the ease (or difficulty) of a substance in bonding to an enzyme's reaction centre. The presence of competing substances is not necessarily a disadvantage. Cells often regulate the pace of an enzymic reaction by composing and channeling the appropriate amount of an antagonist to the enzyme's reaction centre so that the number of "open" keyholes is optimal. Chemists, also, have dedicated long hours to composing the right antagonists, which people take in the form of medicine when one of their enzymes overreacts.

Simple enzymology lessons

Keyholes, keys, and master keys

When keyholes lock safely

Rubisco is strange not because there is an antagonist against the reaction, but because of the nature of this antagonist. Remember from previous pages that Rubisco catalyses the incorporation of atmospheric CO_2 into a pentose (a sugar with five carbon atoms), leading to the production of other useful sugars.

Rubisco accepts CO_2 as a key and O_2 as a master key. In the former case, photosynthesis occurs, whereas in the latter, photorespiration – release of CO_2 back into the atmosphere – takes place.

Therefore, the keys in this case are CO_2 and the pentose. The antagonist is oxygen, which may take the place of carbon dioxide, acting like a master key. The reaction of oxygen with the pentose is an oxidation, resulting in part of the pentose being released back into the atmosphere in the form of CO_2. Remember that this pentose was the result of photosynthetic bonding of atmospheric CO_2 with the help of Rubisco. Therefore, on one hand Rubisco appears to bond and transform CO_2 to sugars and, on the other, to oxidise the sugars produced and to release CO_2 back into the atmosphere, when it receives O_2 in its reaction centre. Is this a case of a futile cycle, of bad reaction centre design, or of a bona fide efficient cause? It should be noted, of course, that the affinity of CO_2 for the reaction centre of Rubisco is much higher than that of O_2. Consequently, although the amount of O_2 in the atmosphere is now 570 times greater than that of CO_2, its high concentration is compensated for by its low specificity for Rubisco's reaction centre. Actually, 20 % of the CO_2 assimilated during photosynthesis escapes back into the atmosphere. Is this a malfunction or a sacrifice offered, and why?

A mistake, a waste, or a choice with a point

Rubisco is an age-old enzyme. It exists in all photosynthetic organisms, even in anaerobic photosynthesising bacteria (those that oxidise H_2S to S), which cannot tolerate O_2. In other words, Rubisco existed even before the appearance of oxygenic photosynthesis – by cyanobacteria, algae, and plants – which oxidises H_2O to O_2. In those early days, the supposedly bad design of Rubisco's reaction centre would not be of any consequence, because there were only traces of O_2, whereas CO_2 levels reached 5 %. Photorespiration, therefore, stood no chance. Even later, after cyanobacteria appeared, the O_2 increase in the atmosphere was slow. Furthermore, for carbon losses to start through photorespiration, not only does O_2 have to increase, but CO_2 must be reduced as well. Both happened in parallel, but extremely slowly. A short time before land was

Photorespiration was not a problem in the primal atmosphere, in which there were not many master keys (oxygen)

colonised by plants (about 450 million years ago), O_2 is believed to have stood at 10 % and CO_2 at 0.5 %. In other words, the O_2/CO_2 ratio was 20, compared with its current value of 570. Problems with carbon losses must have already appeared. Imagine O_2 and CO_2 molecules crowding around Rubisco's reaction centre. CO_2 has much higher access potential, but the probability of access also depends on the population of the particular gases (O_2 and CO_2) that want to enter. It is one thing to have a O_2/CO_2 ratio of 0.002 (as it was during the anaerobic period of the earth's history), but another to have a ratio of 20 (before the appearance of land plants) and quite another to have a ratio of almost 600 (the stabilised value after land plants prevailed). In other words, in less than 80 million years, photosynthesis by land plants managed to increase the O_2/CO_2 ratio up to values comparable to present-day levels.

It often is said that when a trait is a disadvantage, natural selection eradicates or corrects it. Did this actually happen in the case of the defective reaction centre of Rubisco? Studies of the specification coefficient of Rubisco for the two gases involved (O_2 and CO_2) in various categories of algae have shown that indeed there was an attempt at correction. Groups of algae that, according to fossil archives, seem to have appeared more recently, present a much higher specificity for CO_2 than for O_2. In more primordial groups (i.e., those appearing in geologic periods when the O_2/CO_2 ratio was still quite low and, therefore, the problem less intense), the corresponding specification level is lower. However, the level of specification of land plants, regardless of their taxonomic status, remained low and similar to that of their algal ancestors. Therefore, photorespiration seems to be performed at a significant pace in the atmospheric conditions prevailing in the last 400 million years. As a result, land plants throw away part of the carbon they worked to assimilate. If photorespiration were absent, they would have higher growth

In the course of evolution, algae corrected the mistake...

...something plants refused to do...

rates. Why has there been no attempt to correct such imperfections among land plants? Is this a case of indolence and sloth on the part of natural selection, or a case of providence to avoid self-destruction?

...probably for a reason

Uncontrolled photosynthesis, that is, the continuous increase of O_2 and continuous reduction of CO_2 in the atmosphere, may have a devastating impact on vegetation and, consequently, on all life forms on Earth. Increased (exceeding current levels) concentrations of O_2 make plant matter highly flammable, particularly when the temperature is high and there is a lack of water in the atmosphere. Spontaneous fires may become large and catastrophic. There are indications that this happened – for example, during the Carboniferous geologic period ($\approx$ 350 million to 300 million years ago) – and resulted in the coal ores that fuelled the Industrial Revolution. It is estimated that atmospheric oxygen, as a consequence of the intense photosynthesis of the first 100 million years after the earth's colonisation, stood at 25 % to 35 %, the threshold before spontaneous combustion is favoured. Furthermore, the continuous reduction of atmospheric CO_2 and its bonding as organic biomass within animal and plant bodies may lead to its scarcity as the basic food source for plants. It should not be forgotten that the "food" of plants is light (which will be freely available to the earth until the sun collapses, i.e., after 4.5 billion years), water, inorganic elements from the soil (relatively ample), and CO_2 from the atmosphere. The minimum CO_2 concentration a land plant requires to complete its life cycle (i.e., to grow, develop, and have offspring) has been found to be 0.018 %, or half the value of what is available from the atmosphere today. Values close to minimal ones were observed in the recent history of the earth, a few hundred thousand years ago.

Uncontrolled photosynthesis at the planetary level is unfavourable for plants, because oxygen increases and CO_2 decreases beyond measure

This results in plants being burnt by fires and having no food

It seems that plants have learnt their lesson. Uncontrolled photosynthesis and growth lead to a dangerous lack of food (CO_2) and famine. It also leads to the overaccumulation of the plant growth by-product (i.e., pollutant) O_2 and, therefore, an increased risk of fire. It is believed that in the Carboniferous period, fires were widespread phenomena on a planetary scale.

The inevitable conclusion is that Rubisco is a safety valve. By maintaining the age-old design fault, it increases its oxidising function (i.e., the consumption of O_2 and excretion of CO_2) when too much photosynthesis and growth lead to an extremely high O_2 increase and CO_2 decrease. In this way, a forbiddingly high O_2/CO_2 ratio is prevented. When the O_2/CO_2 ratio returns to lower levels, photosynthetic function (i.e., CO_2 assimilation and release of O_2) increases proportionately, as does the growth of plants. In other words, this contributes to homeostasis with regard to the composition of the earth's atmosphere, benefiting not only the plants themselves but also the rest of the earth's organisms as well as the planet itself in its capacity as a huge living organism. It should not be forgotten that even for the rest of the organisms, O_2 *must* remain stable at a level of around 20 %. If, for example, a mountain climber is at an altitude higher than 3,000 to 4,000 m, where the mean oxygen partial pressure is 30 % to 40 % lower, he or she should avoid intense effort to prevent internal hypoxia. If O_2 increases by a similar percentage above its current value, the risk of oxidation damage increases. CO_2 has a less dramatic and immediate effect on animals, but its role as a greenhouse gas in regulating the earth's temperature within limits that make life possible is soundly documented. Therefore, we should be grateful to plants because, since time immemorial, they have been providing us with good-quality air.

Rubisco – a biochemical regulation valve for atmospheric composition

Plants regulate their environment sustainably

Undoubtedly, the double role Rubisco plays makes it one of the most important biomolecules on earth. It has existed for 3.5 billion years, and all the carbon within the bodies of all organisms now and in the past, even those that have become extinct, has passed at least once (but actually many times) through the reaction centre of this enzyme. Moreover, by now it probably is clear that Rubisco is both a detector of atmospheric composition and a regulating valve that intervenes when necessary to correct this composition.

Rampant growth at the expense of sustainability

Although by now the answer is a matter of reasoning, let us return to the question raised earlier: Is photorespiration a malfunction or a sacrifice offered? It seems that plants chose to reduce growth in favour of sustainability. They sacrificed part of their current photosynthetic and growth potential so they may have food in the future and so they may avoid the risk of destruction due to oxidation. In contrast, modern human societies, supposedly made up of rational beings, consider any operation that restricts growth and mitigates temporary gain a violation of individual freedom. Primitive humans, before knowing intellectually that the seed of a plant is destined to germinate and bring forth a new plant, sacrificed part of their crop to Mother Earth (in essence planted, albeit unknowingly) so they would have a crop the following year. Similarly, they did not hunt and kill all the game, even though they were not fully aware of the ecology of populations and the biology of reproduction. Even without being aware of such things, primitive humans acted cautiously, prudently, and wisely, which was compatible not only with their immediate needs but also their future ones. Do their descendants, modern humans – the so-called *Homo sapiens* – behave the same way? One could argue that the social changes during the past centuries (including the scientific revolution) have put humans in a tough position despite their modern prosperity. At least in Western societies, humans are no longer hungry; they do not get physically tired nor are they exposed to natural hazards. They have distanced themselves from nature and its productive processes. They do not get wet, they are not cold or hot, they have not seen a sunrise for years, and sunsets are hidden by the building

What can plants teach us?

block across the street. However, they appreciate the beautiful sunset landscapes framed on the walls of their living rooms and watch documentaries about forests, rivers, valleys, and high mountains. They know from books, not from daily experience, that resources are not infinite; however, they cannot tolerate not acquiring anything offered by industry and promoted by aggressive advertising. Although science has taught modern humans that wasting resources places a burden on future generations, it also has persuaded them that there will always be a "scientific solution" in the end. They believe that because their neighbors have the latest model car, computer, appliance, and so on, they should too. They have thrown away so many things in their lifetime they feel one more will not make a difference. Above all, modern humans have been misled into believing that progress is synonymous with growth and that controlling growth would be catastrophic. Everyone presses for more; everyone promises more. Each person knows (or should know) deep inside that the first thing to collapse, if growth and consumption are reined in, would be our modern economic systems. On the contrary, continuing this current pace of economic growth, energy, and consumption might lead to an ecologic crisis, but not an ecologic collapse. As will be shown later, ecologic crises bring the date of expiration for sensitive biological species closer. It is true that species come and go; none is eternal. If some have proven stronger during ecologic crises and have lived longer through the geologic aeons, these are the organisms we consider to be lower and more conservative, including plants. This is what has been learned from palaeontologic records.

Rubisco's valve is much more instructive than the preceding paragraph indicates. When photorespiration was discovered during the 1960s and 1970s, its molecular biochemical mechanism was determined and it was confirmed that the process in question reduces photosynthesis and, therefore, productivity. The question asked earlier was immediately raised: Was it a case of bad design or of a bona fide efficient cause? Despite reservations among "conservative" supporters of the latter possibility, the "bad design" and evolutionary fault views prevailed within the scientific community. Supporters of these views convinced funding organizations that the bad design could be corrected through human intervention, using the rationale that organisms may be mutated to benefit us, that we can intervene and correct what nature allegedly did not achieve in past geologic aeons. Huge amounts of money were wasted to fulfill biotechnologic promises that photorespiration would be eradicated and, consequently, plant productivity increased. *Homo sapiens*, the capstone of biological organisms, undertook to correct the mistakes nature had made. In this case, as in many others, it was proven that all the biotechnology sorcerer's apprentices and their naïve financiers were wrong. Soon it was discovered that those who claimed there must have been an efficient cause were right. In the mid-1980s, photorespiration was found to play an important role in protecting the photosynthetic apparatus against certain distressing conditions (e.g., a combination of water scarcity and high light intensity). Moreover, at the end of the 1990s, convincing evidence began accumulating with regard to the planetary regulatory effect of photorespiration on atmospheric composition.

Why should one not pretend to be a sorcerer's apprentice?

Size and Age are not Always Related

Back to Methuselahs

With regard to the Methuselahs of the plant kingdom, a question may be raised: Are the eldest plant organisms also the largest, that is, is size correlated with age? One reasonably might assume the answer is yes. Usually, age-old plants are huge trees. However, before the question is answered, the reader should remember two fundamental characteristics of plant growth, which were described in the first pages of this book. The size of a plant is not stable. A pine might remain a bonsai tree or reach a height of tens of meters and a weight of several tonnes. Furthermore, many plants are capable of asexual reproduction (e.g., root suckers), whereupon their descendants are not separated from the mother plant, thus creating numerous clones. Clones may create a network (above or, in most cases, underground), which means that every descendant becomes the mother plant of the next descendant. In every series, all descendants are connected with one another as well as to the mother plant. Obviously, the whole system is a clone-individual developing radially, symmetrically, or asymmetrically, gaining territory and exploiting the resources within it.

Age-old giants...

Many small creepers in arctic and alpine regions are clones. As was already mentioned, a typical feature of these regions is their brief summers, during which the temperature rises somewhat for only few weeks, within which plants may grow. What is the growth rate of a plant, including the clones, under alpine conditions? Some scientists were patient enough to study this over many years in the alpine dwarf plants *Carex curvula* and *Empetrum nigrum*. Accounting for all the necessary parameters and performing all the necessary measurements, they concluded that the mean pace of radial development is around 0.3 mm per year. Therefore, if the radius of a clonal plant is determined, its age may be estimated. Yet, what is the radius of such a plant? This is where the puzzle starts. Because space is not infinite, growth is interrupted or decelerated, or it changes direction if the growing clone encounters an obstacle. Usually, the obstacle is another individual of the same species, that is, the competition among clones for vital territory plays a decisive role. The close proximity of

...and age-old dwarfs

Short alpine summers offer few opportunities for growth, which is so limited...

individuals often makes it impossible to distinguish among them, and the problem of the boundaries of each individual cannot be resolved with the naked eye. Spatial identification of particular individuals using molecular DNA techniques has shown that their maximum surface area is around 1 m^2. Therefore, the radius of the system is about 0.6 m. However, each clone, from the centre of the system (where the mother plant happened to germinate) to its periphery follows a dog-legged path. Therefore, one may conclude that the length of each clone (which is the result of the successive establishment of root suckers) is double, that is, 1.2 m, which is added on through an annual growth rate of 0.3 mm. It is as if one were asked to cover the distance from one end of his or her office to the other using match heads, placing one next to the previous one every year. It would take 4,000 years to cover the distance; yet, this is the minimum estimated age of these humble (in terms of form and size) plants that hardly anyone pays attention to.

...that even small, hardly noticeable bushes may be centuries old

Although these plants flower from time to time, the seeds produced do not germinate because they fall on the dense grid of clone stems, which leaves very few interstices to provide access to soil and humidity. Indeed, rebirth (i.e., the appearance of new individuals) is extremely rare. Existing elderly plants "were born" shortly after the end of the last ice age, without any descendants through sexual reproduction; they have produced only clones. Functionally, these plants are immortal.

Practically immortal plants...

Generally speaking, at high altitudes there is a tendency for older ages, even in normally short-living plants. Some semelparous species, which thrive along an extensive altitude zone, complete their life cycle and die within 2 years at lower altitudes; however, they live twice as long at higher elevations. The low growth rate imposed on plants in such environments makes them particularly vulnerable to disruption, because any destruction is difficult to repair. If the overall conservation of habitats is the state's duty, each of us can contribute privately by walking only along footpaths. A mere step outside the path might destroy century-long efforts.

...if they are stepped on, centuries of efforts are destroyed

How Tall can a Tree Become?

The world champion in height is the conifer Sequoia gigantea: it is as tall as a wind generator and as heavy as two blue whales

Although a plant's height does not always reflect its age (consider the age-old creeping branches of alpine plants), among trees, size (height) is correlated, albeit loosely, with age. Undoubtedly, the best performers are the huge conifers of the *Sequoia gigantea* species in California. The tallest ones reach a height of 110 m (as high as a 30-storey skyscraper or a tall wind generator) and a weight of 200 tonnes (the biggest dinosaurs and largest whales are dwarfed by them), and their age is estimated at around 2,000 years. For record-keeping purposes, suppose the tallest tree today stands at 113 metres whereas several individuals within the population have reached a height of 110 m. What stops trees from growing higher?

What prevents trees from growing even taller?

Apparently, the problem of mechanical support for this great weight is not a critical factor. The mechanical properties of xylem suffice not only to keep these trees standing, but also to protect them from falling, even under gale-force winds.

Theoretic models and calculations indicate that the problem is hydraulic in nature: the water reaching the uppermost leaves of the canopy is not enough for growth. This was proven recently when a team of scientists climbed to the top of some of these giants. Equipped with the necessary devices, the scientists measured typical physiologic parameters (e.g., photosynthesis, leaf water loss and content) at regular intervals. They concluded that although there was

Why do the uppermost leaves of a tree suffer from water shortage?

enough water in the soil, the uppermost leaves suffered from a water shortage. Therefore, the initial question may be modified as follows: How do plants raise water from their roots to their uppermost parts? The answer provides an opportunity to describe, in simple terms, the movement of water in the plant's body, which – as already explained – is not equipped with a central pump similar to that of animals (heart).

Some points to be raised are:

- What are the water entry points at the root level?
- What is the course water follows to reach the rest of the plant, and through which tissues and organs?
- What are the water exit points towards the atmosphere?
- What is the driving force moving water against the force of gravity?

Water spontaneously moves from higher to lower levels. The opposite requires energy consumption.

Water moves spontaneously from higher to lower levels, and during this movement work is carried out. If a dam is constructed on a river, the controlled waterfall may be exploited by transforming its kinetic energy into electricity. If, on the other hand, a mass of water must be raised against gravity, the necessary energy must be provided. In principle, the same thing needs to happen when water flows up a plant's body, from the roots to its uppermost parts. To bring water up from a well, a pump, running on oil or electricity, is needed to provide the necessary work. The pump may be a lift or a suction pump, that is, one that pushes water from below or one that pulls water up. Plants use both pumping methods in succession. The root works as a lift pump at night and the leaves as a suction pump during the day. There is one difference, however: the plant uses the chemical energy it consumes at the roots wisely so it can perform other functions as well; therefore, the same quantity of energy pushes water up at no additional cost. Furthermore, the plant cleverly exploits the natural

Suction and lift pumps

properties of water so that the operation of the leaves' suction pump entails no additional energy cost to the plant.

If a root is dug up carefully and its tapered end observed under a microscope, a zone of tiny hairs – a few millimeters from the tip – may be discerned. The entry of water and other mineral nutrients the plant needs occurs mainly through these *root hairs*. The hairs drastically increase the area of the root available for absorption in relation to the root's volume, thus helping water and mineral nutrients to enter. Yet, although water enters the cells almost freely, the same is not true for the essential inorganic nutrients. There are more than 200 elements in the soil, usually in the form of ions. Only 17 of them are necessary for plants. The rest not only are unnecessary, some are also toxic to all organisms, including plants. Examples of toxic elements are the so-called heavy metals (mercury, lead, etc.). Furthermore, not all 17 necessary elements are required in the same quantities. For example, magnesium, calcium, potassium, nitrogen, and other elements are required in large quantities whereas only traces of copper, iron, manganese (and some others) are necessary, and absorption of the latter group in larger quantities is harmful to plants, although small quantities are essential. Therefore, not only must the entry of nutrients be controlled, but their rate of entry must be appropriate.

A plant's water entry points are its tiny root hairs

The water entering at the root contains useful, useless, and toxic chemical elements

The points controlling the entry of inorganic elements are located on the membrane surrounding plant cells. Special proteins that permeate the cellular membrane allow or block and regulate the rate of entry. Actually, every prospective visitor must show its passport, that is, its structure. The protein (point of entry) has a domain (reaction centre) exposed to the external side of the membrane, and its structure is such that it attracts only one element (or a few elements) of the right structure. If the visitor is not "registered" as

Before entering a root, every chemical element must show its passport, that is, its structure

undesirable, the gate (imagine it as a tube) opens and carries the visitor to the cell's interior. After they enter the cell, the visitors, through corridors interconnecting the root's cells – arrive at the periphery of a central cylinder inside the root. The protein's structural changes (necessary for the gate to open) consume energy, which is produced from the cellular respiration of sugars, originating from the process of photosynthesis and transported from the leaves to the roots.

Border control is costly

The root's central cylinder contains transportation vessels (similar to arteries in animals). These vessels have ligneous walls that begin at the roots, go through the stem and branches, and end at the leaves. Vessels are visible as hard, less-green branching "nerves" on leaf laminas. At the root, at the base of the vessels, the required elements entering are accumulated and pushed together at high concentrations, limiting the space occupied by water. As a result, water from adjacent cells tends to enter at the base of the vessels, locally increasing pressure, which has to be "relieved." To better understand the course of events, it must be said that:

When welcomed, chemical elements cross into the interior of the root vessels; they carry along water as well

- Pressure cannot be relieved by pushing entering elements back to the soil. In fact, crossing the entry gate of the cell membrane is a point of no return. The nutrients are essential and cannot be sent away.
- Pressure cannot be "relieved" by increasing vessel volume. Vessel walls are ligneous and cannot stretch.
- A local reduction in water concentration at the base of the vessels (because water is displaced by the mineral nutrients entering) leads to an inflow of water from the adjacent cells (and ultimately the soil), thus further increasing pressure. This water flow is the result of the law of diffusion, which leads to the spontaneous movement of water from a site of higher concentration (the cells around the vessels) to one of lower concentration (within the

vessels). This force overcomes the fact that the pressure is greater inside the vessels than in adjacent cells. When a substance crosses a selectively permeable membrane (such as those of the cells), both pressure and diffusion forces are involved; this phenomenon is called *osmosis*. This issue is rather technical and beyond the scope of this book.

Thus, pressure increases at the base of the vessels and water is pushed upwards

The final result is that the water pressure at the base of root vessels is relieved following the path of least resistance – that is, upwards – carrying water, along with nutrients, to plant parts above ground. In other words, the root pushes the water upwards, against the direction of gravity, using the pressure built at the base of the vessels as the driving force. The necessary energy for this push is, ultimately, the chemical energy consumed in the root cells to transport the necessary inorganic nutrients from the soil to the interior of the cells. In other words, it is energy that would be consumed anyway to transport the mineral nutrients needed to feed the plant.

The root, therefore, operates like a lift pump. Because the cost to transport the chemical elements has already been paid, the upward pumping of water entails no additional cost.

At night, the lift root pressure recharges the plant body, replacing daily losses, because plants lose water during the daytime in quantities exceeding those that can be replaced by movement through the root. In daylight, suction pumping prevails because of leaf activity.

During the day, leaves photosynthesise. In the chloroplasts, the subcellular organelles within leaf cells, special pigments (chlorophylls) absorb solar energy and, through a complex redox process, transform it to chemical energy and reducing power, which are then used to assimilate carbon dioxide (CO_2) and transform it into sugars. The source of CO_2 is the atmosphere. However, the wax-like top layer of the leaf epidermis (the cuticle) is not permeable to CO_2. Therefore, land plants have acquired cellular systems on leaf surfaces (guard cells) that form valves of various openings called *stomata*. The plant adjusts the degree of valve opening according to its needs. At night (when photosynthesis cannot take place because there is no light and, therefore, no CO_2 entry is necessary), the stomata are shut. Opening of the valves during the day allows CO_2 to enter, but inevitably, it also allows water to escape in the form of vapour. The phenomenon of water loss from the leaves in the form of vapour is called *transpiration*. Transpiration is what shifts the centre of the force driving the water's rise from the root (during the night) to the leaves (during the day) and transforms the pump from a lift to a suction pump. How does this happen?

In daytime, a second pump is set to operate in the leaves, but this is a suction pump

According to Fick's law of diffusion, a fundamental law of physics, any substance moves spontaneously from where its concentrations are high to where its concentrations are low. It is diffusion that brings a flower's perfume or exhaust fumes to the olfactory receptors in one's nose. The same law also ensures that the rate of movement of the substance being diffused between two points is proportionate to the concentration gradient between those points. In other words, for a given distance between nose and flower, the perfume will arrive at the nostrils faster if essential oil production is higher at the source. In the leaf's case, the gradient in water vapour pressure between the internal and external atmospheres is huge. The interior of the leaf comprises cells whose surfaces border on adjacent cells, but their greatest part is exposed to the leaf's internal atmosphere. They are like the apartments in a block of flats with many commonly used corridors. Depending on the plant, the interior gas spaces (those free of cells) within a leaf may range from 2 % to 50 % of the total leaf volume. In other words, to a great extent, the leaf is internally hollow.

Leaves are not solid inside

Cell walls, with their mesh of cellulose fibrils, are like tiny wicks sucking water from the vessels (the "nerves" of the leaves). The water either feeds the cell's interior or evaporates into the internal atmosphere. This atmosphere is suffocating as far as humidity is concerned: the humidity is 100 %, that is, the air is saturated with water vapour. In fact, the internal atmosphere is always within a cloud. Because the external atmosphere normally is much less humid, there is a permanent, strong tendency for leaves to lose water vapour during the day, when the stomata are open. The quantities of water escaping to the atmosphere are enormous. A small tree, equal in height (and weight) to a human, might lose between 6 and 30 kg of water per day. To maintain its good health, the tree must replace this water from the soil. *Mutatis mutandis,* a human of the same height and weight would need 24 to 120 glasses of water per day. In small grassy plants, water losses are even more impressive. In a single day, a corn plant transports from the soil to the atmosphere an amount of water equal to three times its weight. This is not necessarily a reckless waste. When water evaporates, it removes heat from the surface it escapes (this heat is called *latent evaporation heat*), thus cooling the leaves.

The internal leaf atmosphere is always saturated with water vapour, like a Turkish bath

If humans lost as much water as plants, they would need to drink up to 30 kilos of water per day

For a water molecule to escape from the aqueous to the gaseous phase, it needs to break the so-called hydrogen bonds formed between water molecules. Hydrogen bonds are weak electric bonds created by the bipolarity of the molecule. Although there is no net electric charge, there are slightly negatively charged oxygen atoms and slightly positively charged hydrogen atoms asymmetrically distributed in space. Therefore, water appears as a dipole and consequently its molecules form a characteristic, strong grid because of the electrical attraction between the two opposite molecule charges. Hydrogen bonds are responsible for a series of important properties of water that make it absolutely essential for life. This is why efforts to detect extraterrestrial life focus on planets and satellites that likely contain water. Water properties will be revisited in this text as

The physicochemical properties that make water essential for life are attributed to its bipolar character

necessary. Regarding latent evaporation heat (which cools leaves), the following may be stated: During evaporation, the hydrogen bonds restored among adjacent water molecules need to be broken. The energy needed for this is quite high, and it is always provided by the water mass itself, which cools down further the more it evaporates. Evaporation may be accelerated if external energy becomes available, that is, in the form of heat. This is why drying clothes become cooler, the washing dries faster in the sun, and perspiration relieves us in the summer.

Water evaporation cools leaves

Water evaporation from the leaf's interior into the atmosphere through the stomata creates a water deficit that needs to be restored. The walls of the leaf cells act like a wick and absorb water from the vessels. The suction pump is ready, but from how great a depth can such a pump draw?

With regard to pumping at the molecular level, it is not very difficult. A water molecule attracted by the pump will draw upwards an adjacent water molecule with which it has restored hydrogen bonds. The adjacent molecule attracts a third molecule, which then attracts a fourth one, and so on. Of course, the force of gravity resists this movement. Because the two (attraction upwards, gravity downwards) are counterforces, the mean distance between water molecules tends to increase. The column of water stretches like rubber; as the distance between the molecules increases, their density and pressure decrease. If the two counterforces increase too much, there is a risk of the hydrogen bonds breaking. Then, the molecules of water that are relatively distant from the pump will succumb to the weight of the water column and collapse to the bottom of the tube. Once the water column has broken, pumping is interrupted.

The force of the suction pump against the force of gravity

Because these are counterforces, the water columns in the xylem vessels are at risk of fracture

The particular structure of water allows it to form strong hydrogen bonds between its molecules, that is, to increase its cohesion. Water columns are not broken as easily as those of other liquids, which are not as easy to pump as water. For the same reasons, water molecules not only present high cohesion to one another, but they also have a great affinity for the xylem vessel walls. Therefore, water columns within the vessels are impressively resistant to tension, although in essence they are "hanging" from leaves tens of meters high. The probability of fracture increases when the counterforces increase too much. The transpiration force pulling upwards is quite dependent on environmental factors, such as atmospheric humidity and temperature (remember that the washing dries faster when the air is dry and the temperature high). Plants, however, recognise the danger and adjust the opening of the stomata to avoid the risk of dehydration and embolism (i.e., interruptions in the continuity of the water column in the vessels). Of course, when the opening of the stomata decreases, photosynthesis decreases proportionately, because CO_2 entry becomes more difficult. Therefore, it is easy to understand the importance of adjusting the stomata-valves in response to opportunities and needs. When the risk of dehydration and embolism prevails, the stomata close, because survival is more important than growth. When normal conditions are restored and the risks removed, the stomata open and provide an opportunity for photosynthesis and growth. This is a fine balance, based on sound risk and opportunity assessment, that is, the accurate perception of environmental conditions.

However, water tolerates tension quite well

When the counterforces increase too much, the stomata close to limit the power of the suction pump

If the plant can adjust the transpiration suction pump for short periods, the force of gravity is more stable in character: it depends on the weight of the water column, that is, on the dimensions of the vessels – their length and diameter.

The diameter of vessels must have been quite a headache for natural selection; the mean diameter of around 150 μm (millionths of a metre) reflects the final compromise between the needs of plants and the laws of

Why do plant vessels have a diameter of about 150 μm?

physics as applied to fluids. A large diameter ensures high flow speed (for a given driving force) because of the reduction in friction. However, it also increases the weight of the water column and, therefore, the risk of fracture and embolism. A small diameter increases friction but favours capillary phenomena and reduces the risk of embolism. For example, imagine drinking a liquid through a straw. What would happen if the straw were ten times wider or narrower? Deviations from the mean diameter of 150 μm in various plant species are related to the climate of the habitats in which these species thrive. If environmental conditions do not predispose for embolism, vessels may be wider.

Given that the vessel diameter cannot deviate significantly above or below 150 μm, it may be concluded that the weight of the column is determined mainly by its length. Theoretic consideration of the strength of a water column with a diameter of 150 μm, which "simply hangs" from various heights, has shown that the mean height permissible is 130 m. *Simply hangs* means that transpiration is minimal, the stomata are half-shut, and the force of attraction is low. Obviously, increased transpiration reduces the maximum length allowed.

The maximum height of trees is determined by the properties of water and the laws of physics. This height cannot exceed 130 m

The results of this theoretic consideration were confirmed by the climber-researchers. The uppermost leaves of the tree species *S. gigantea* were small, thick, and hard, with low water content and low rates of transpiration and photosynthesis. All their features indicated that they were suffering from water shortage. The limits of tree heights, therefore, are imposed by the laws of physics.

There are other reasons it is so difficult for trees to be that tall, or even a bit taller than their neighbours. When all trees are the same height, they help each other face the wind, which is stronger as the altitude rises. The wind may scathe sensitive (mainly developing and therefore soft) plant tissues either as a result of collision with the particles (dust, small icicles) it contains or because of branches colliding with each other. As in human communities, the taller a tree grows, leaving behind the anonymous safety of a group defence, the more problems it has to face and, indeed, face alone.

Embolism frequently occurs in plants. . ..

In conclusion, it must be stressed that contrary to humans, among whom embolisms are rare but dangerous, among plants embolisms are common (maybe even a daily phenomenon) but not dangerous and almost fully curable. It might be worth examining why this is so – this is another difference between plants and animals.

As was mentioned earlier, the vessels transporting water are renewed once a year. Macroscopically, they form the so-called rings in the tree trunk, which are used to estimate the age of trees and the climate of corresponding time periods. Therefore, even total incapacitation of vessels because of embolism may affect the plant until the next growth period, at the latest, because fresh vessels are then recreated.

. . .however, it is not dangerous. . .

Vessels originate from the permanently embryonic cells of the cambium, which create a cylinder underneath the bark and on the outside of the xylem. Cells destined to become vascular elements through the differentiation process are laid along a vertical axis. Then, the cells undergo programmed cell death, whereupon their organelles and membranes are decomposed while their cell wall becomes stronger and its sides are lignified along the axis. On the sides perpendicular to the axis, the wall appears perforated, creating grids or other forms of communication channels between successive vessel members. Therefore, although the continuity of a vessel is, in effect, undisturbed, allowing the upward movement of water, the traces of the initial cells are marked by the separating grids. Furthermore, at selected points on the walls parallel to the axis, reinforcement and lignification of the wall do not take place, leaving a thinner wall and forming a pit. Similarly, another pit is created in the adjacent vessel member at a corresponding point so that the pits form a set.

Where water flow is interrupted, the water is replaced by a low-pressure air bubble. This is air dissolved in the same water not air from the atmosphere, as there is no direct contact between the vessels and the atmosphere. The bubble increases and occupies the space of a vessel member, but the separating grids do not allow it to expand along the water column. This provides some relief, particularly to the higher parts of the affected vessel because they no longer have to support the weight of the whole column. Of course, the flow of water in the specific vessel stops, and this is a problem. However, the water ascending in this vessel can use the detouring path of the pits as the region of the side walls that present lower resistance against water flow. Therefore, part of the flow is restored and the problem is alleviated somewhat. It is clear that the vessel's design provides for the possibility of a bypass. In similar situations among humans, the bypass would be undertaken by a surgeon's intervention. Finally, the bubble disappears the following night, when the stomata of the leaves close (i.e., the suction pump of transpiration stops functioning) and the lift pump of the root prevails. Embolisms in plants might be annoying, but they are not a serious risk for their sound health because of their temporary character, the possibility of a spontaneous bypass, and the repair of the lesion when the root pressure increases at night.

...because the possibility of spontaneous bypass is an integral part of the vascular structure

...and the repair of damage takes place easily the following night, when the root lift pump starts operating again

Life Span and Species Immortality

Species appear, and at some point they – inevitably – become extinct

If certain plants are the oldest living organisms on Earth, is the same true about plant species? We know that species appear, originating from other species, and they evolve until their form changes so much that they are considered a new species. Of course, these changes occur not only in the form of plants but in their physiology as well. One can only guess about the physiology of extinct species; however, their form may be reconstructed quite accurately from their fossilised remains. Therefore, although the classification of existing organisms into various species is based on morphology, chemistry, and, recently, genetic material analysis, that of extinct species is based exclusively on their form. Although this entails some uncertainty, no other method exists. To clarify further, one may reformulate the question as follows: How much time passes from the appearance of a species until its extinction? It also should be stressed that although appearance presupposes evolution from a previously existing species, extinction does not necessarily lead to the appearance of a new species. A species may arrive at an evolutionary dead end, although in most cases its genetic material is perpetuated, although slightly modified, in some successive species.

How much time passes from the appearance of a species until its extinction?

Therefore, what is the mean life span of species? Naturally, it depends on the specific organism. In the animal kingdom, the champions of longevity are marine foraminifera (small protozoa with shells), the species of which have been present for an average of 25 million years. The second oldest are bivalves (e.g., oysters) and gastropods (e.g., snails), with a mean life span of 12 million years. In the rest of the animal kingdom, particularly among groups appearing later in the earth's history, the mean species life span is less than 3 million years. In the case of mammal species, the average life span is estimated at 1.5 million years. Of course, these are mean values for each group, within which there may be successful and unsuccessful cases showing significant deviations.

Among plants, species longevity is completely different. Among small bryophytes, species life duration exceeds 20 million years. Although herb species are not the oldest, they live longer than mammals, up to 3.5 million years. The world champions are the conifers and

The mean life span of plant species is far longer than that of animals

The stately Gingko biloba and the humble *Equisetum* (horsetail) have remained almost unaltered for 300 million years

the primordial phyla Gnetophyta, Cycadophyta, and Ginkgophyta, species of which have a mean life span of 54 million years. Of course, among these species are deviations in both directions. At the genus level, impressive cases are those of *Sequoia*, *Araucaria*, plane, and walnut trees, at 100 million, 150 million, 120 million, and 90 million years, respectively. Even more impressive are the small-plant genera *Lycopodium* and *Equisetum* (known as horsetail or snake grass), which have not changed their form significantly in the past 300 million years. At the species level, *Gingko biloba*, the only living representative of the Gingkophyta that used to dominate the earth 200 million to 100 million years ago, has remained impressively unchanged to date. *Gingko biloba* is a beautiful tree used to decorate streets, squares, and parks throughout central Europe. In the wild, it is found only in a few remote regions of China. The stately gingko and the delightful equisetum excite naturalists because they are as recognisable to modern observers as they were to someone travelling in the past, some hundreds of million years ago.

What is it that makes these plants so successful through aeons? What are the properties that allowed them to exist for such long periods without any change, to show such stability of form, and to survive climatic changes and planetary disasters? Why is it that the awesome dinosaurs became extinct while the fragile equisetum has survived? These and many other questions are explored in the chapter on the evolution and history of plants on planet Earth. This chapter shows that mass extinctions of animal species, which have marked the earth's history at least five times, are not accompanied by similar extinctions of plant species. In a hecatomb allegedly caused by an asteroid colliding with the earth 65 million years ago, 60 % to 80 % of animal species (including the dinosaurs) disappeared. Although there was a corresponding slight recession of vegetation in some regions (probably due to extensive fires), soon all was restored and the floras (i.e., the sum of all plant species) returned almost to their previous state. The next chapter also shows that a combination of strategies and properties allows plants not only to face severe ecologic traumas (such as those causing mass extinctions of animal species), but also to increase in numbers and dominate the earth with characteristic steadiness, determination, and persistence.

Genetic stability and resistance against ecologic trauma

Chapter 4
Short Evolutionary History of Plants

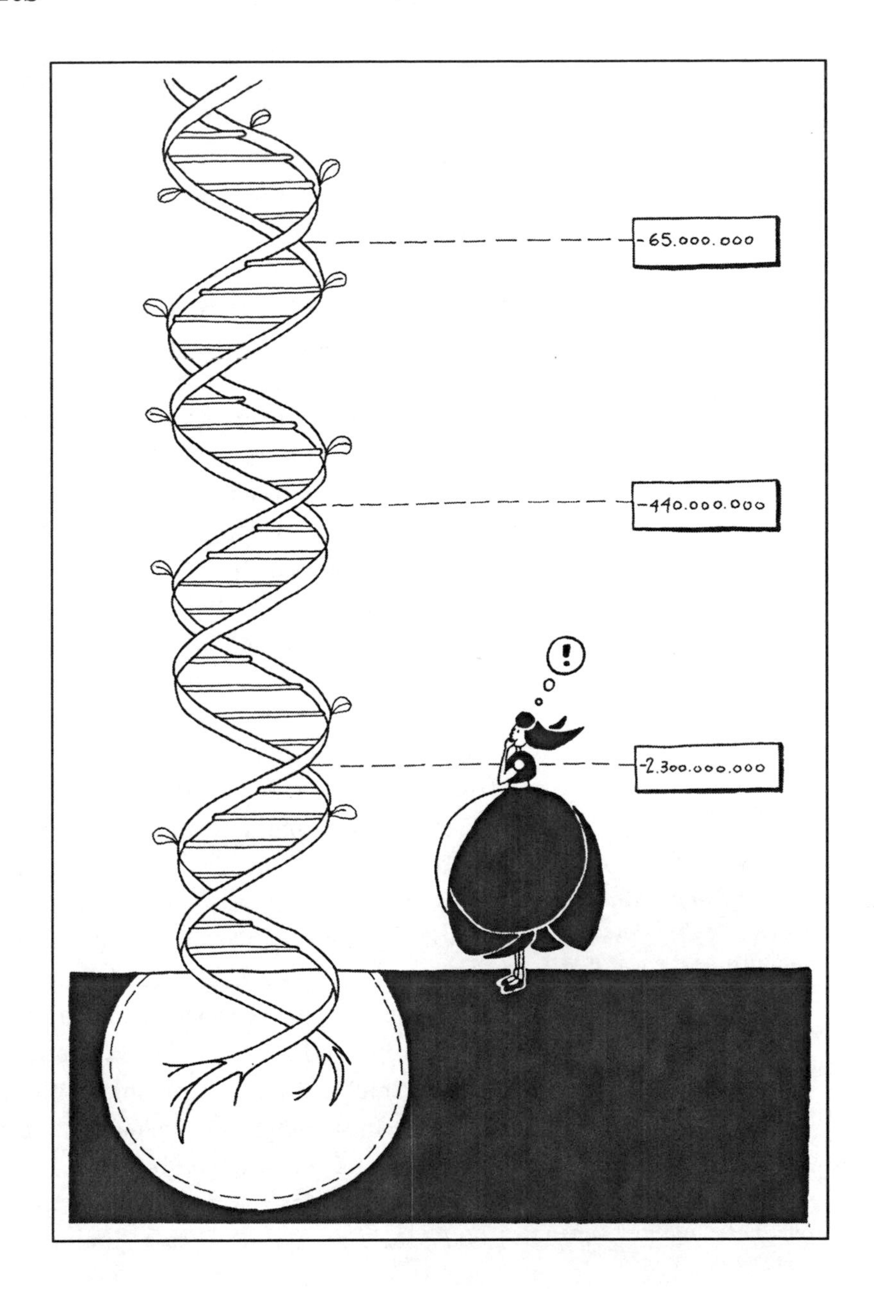

Which Organisms are Characterised as Plants?

So far, it has been assumed that we know which organisms are considered plants. However, this is a question that cannot be evaded, and perhaps this is the right moment to answer it. There is no dispute as to whether a plane tree, for instance, is a plant. However, what is the critical feature that leads to classifying a plane tree in the plant kingdom? It is not the presence of chlorophyll and photosynthesis, because that would exclude the parasitic plants that develop suckers inside other plants to absorb what they need. Hence, they do not need to photosynthesise and have no chlorophylls. Yet, parasitic plants have all other plant features; furthermore, they have evolved from green and photosynthesising ancestors, which at some point found recourse to a parasitic way of life. Additionally, the presence of chlorophyll and photosynthesis is not a sufficient criterion, because we already know that there are photosynthetic bacteria that contain chlorophylls. In the past, the prevailing view was to consider even fungi as plants because they have a cellular wall; however, today they are considered a separate kingdom.

The presence of chlorophyll and photosynthesis is not a sufficient criterion for the classification of an organism as a plant

Are algae plants? Considering their overall biology, morphology, behaviour, and nonmotility, as well as the fact that they photosynthesise and have generally the same nutritious needs, the answer may well be affirmative. However, modern molecular biology has shown that algae cognate more with amoebas and the protozoans than with plants. Therefore, today a classification of five kingdoms is accepted: Monera (bacteria), Protista (algae and several other heterotrophic, unicellular, and multicellular organisms with fundamental differences from other organisms), fungi, plants, and animals. Plants include bryophytes (mosses), with about 16,000 species; pteridophytes (ferns), with about 11,000 species; gymnosperms (mainly conifers and

Molecular biology has shown that algae are more closely related to protozoans than to plants

A brief description of the four major plant groups

some other smaller groups), with about 800 species, and angiosperms, or flowering plants, with about 235,000 species.

Bryophytes are small plants that prefer wet habitats; this makes transport tissues (vessels) redundant for the transport of water and photosynthetic products. Pteridophytes are larger and have vessels. However, in both species, reproduction requires a particularly wet environment because their motile (usually with a flagellum) male reproductive cells (gametes) must swim to reach their immotile female ova to fertilise them. Although several pteridophyte species can live in dry environments, their reproduction always requires water in a liquid state. *Mutatis mutandis*, these plants have sex in the rain. On the contrary, in gymnosperms and angiosperms, male reproductive cells (pollen grains) cannot move on their own but are equipped with adaptive mechanisms that prevent dehydration. These cells are carried to the female cells by the wind or through the intervention of insects, birds, or bats. Therefore, the reproduction of gymnosperms and angiosperms (flowering plants) is better adapted to dry environments, and it has liberated itself from the need for liquid water. The final result of the fusion of the reproductive cells is to create forms suitable for dissemination; in the bryophytes and pteridophytes, these are unicellular spores, whereas among gymnosperms and angiosperms, they are multicellular seeds. The seed includes the embryo, as well as nutrient storage tissue, to help the growing seedling with necessary structural blocks and energy. The seeds of angiosperms are better protected than those of gymnosperms because of the development of ovarian cells around them so as to form a fruit. The flowers of most angiosperms are constructed in a way that helps pollination through the intervention of animals (mainly insects and birds), whereas the corresponding forms (cones) of the gymnosperms serve pollination with the help of the wind. Maybe this

Bryophytes are always small and have no vessels

The male gametes of bryophytes and pteridophytes move to the ova with the help of flagella, similar to those of human spermatozoa

In gymnosperms (conifers) and angiosperms (flowering plants), male gametes are transported to the ova by the wind or through the intervention of insects

Bryophytes and pteridophytes prefer to reproduce in the rain. The rest of the plants can also manage in droughts.

is why angiosperms (there are 235,000 species of flowering plants) have had such an incredible evolutionary course compared with gymnosperms (only 800 species). Flowers advertise their presence with bright colours and enticing odours, so pollinators can find them against the green background of leafage. Furthermore, they reward their pollinators by providing them with pollen and nectar, high-quality nutrients with a high energy content. This is how such plant species ensure they receive visitors to transport their pollen to their mating partners. It is a win-win situation for all involved, as well as being beneficial for natural balance. Of course, this also means that the reproduction of such plants is not a random affair. An insect that finds a flower it prefers will then visit the corresponding flower of a neighbouring plant, hoping to be rewarded in the same manner. In contrast, the reproduction of gymnosperms depends on the whims of the wind. Of course, in either case, plants have no idea who their actual sexual partner was – more about this in Chapter 5.

The history of plants forms the basis of the history of life on land

Therefore, the plant story presented here is the story of the aforementioned groups – bryophytes, pteridophytes, gymnosperms, and angiosperms. Quite often, these groups are referred to as *superior plants*; however, this name does not imply higher quality. Simply put, these plants are better adapted to life on land as compared with their ancestors, freshwater green algae (chlorophyceae). The few plant species that now live and complete their life cycle underwater are former land plants that returned to an aquatic existence, resulting in the corresponding recession of the features that make living on land successful. A familiar case is that of *Zostera marina*, a sea grass (a relative of wild cereal ancestors) that creates extensive submarine meadows in sandy, shallow waters; its leaf residues lie on expansive stretches on the shore after a storm with strong waves.

Life on Planet Earth before the Appearance of Plants: Some Key Biological Episodes that Paved the Way

As already mentioned, life probably started in the sea. By the lower Devonian period, the geologic period to which the first land plant fossils date back – 410 million to 420 million years ago – all algae groups, as well as most of the metabolic pathways known to us, had already evolved. Of course, before land could be colonised, it had to exist. It seems the solid crust of the earth started forming 4.2 billion years ago, a mere 400 million years after the planet was formed. The first gigantic continent probably emerged 1.9 billion years ago, whereas life in the oceans already existed 3.7 billion years ago. However, the first terrestrial environments were rather inhospitable: the mean temperature was between 30 °C and 50 °C (i.e., double or triple what it is today), and the O_2 concentration was no more than 1 % (i.e., 21 times less than today). Because of the lack of O_2, it was not possible for the ozone layer to form in the stratosphere. Therefore, terrestrial environments were bombarded by harmful high-intensity ultraviolet radiation, making it impossible for land to be colonised. Besides, high levels of CO_2 and CH_4 (methane) gases in the atmosphere created an intense greenhouse effect. The first oceans were formed around 4 billion years ago, but these were equally inhospitable, with temperatures close to the boiling point of water (80 °C to 100 °C) and a high level of acidity. The gradual cooling of the oceans and a reduction in their acidity allowed microscopic anaerobic bacteria to appear 3.7 billion years ago; some of these were capable of photosynthesis.

The first sufficiently large land surface emerged 2 billion years ago

However, it was inhospitable, with an intense greenhouse effect, minimal oxygen, and no stratospheric ozone layer shield

Life in the form of bacteria existed in water masses as early as 1.7 billion years before the land emerged

The bacteria known today present a wide range of metabolic pathways and can break down a multitude of organic and inorganic substances, virtually everything around them. In other words, depending on the bacteria group under investigation, they are capable of "eating" anything, even materials they did not encounter in their evolutionary past –manmade materials, such as plastics and car tyres – while they biosynthesise a wide variety of organic substances with the same ease. There is nothing to suggest that their distant ancestors were any less capable. If this is true, it can be deduced that the basic metabolic pathways, including the synthesis of nucleic acids (DNA, RNA) and proteins, already existed in the first stages of life on Earth, a mere few million years after the temperature of the oceans became compatible with the presence of living organisms. Some of these biochemical capabilities were inherited by their descendants, that is, present-day unicellular and multicellular eukaryotic organisms (fungi, protists, plants, and animals).

The basic metabolic pathways developed in bacteria impressively early, within a few million years. Many of these pathways have been inherited by multicellular organisms almost totally unaltered.

Bacteria are tiny unicellular organisms with no apparent internal cellular organelles. Their internal metabolism must be organised in "compartments" of specialisation. Some processes take place within membranes and others in the watery cytoplasm, in which there must be "sites" – albeit without clear boundaries – of some biochemical specialisation. This is necessary so that substrates and products of one metabolic pathway are not confused with those of another. This type of cell is called *prokaryotic*, as opposed to *eukaryotic* cells, which are equipped with apparent cellular organelles, visible under a microscope, each with a separate function. For example, the nucleus contains the chromosomes, which contain the DNA, that is, the organism's growth and development data and programming. The mitochondria perform respiration, the chloroplasts perform photosynthesis, the vacuoles are storage sites of plant cells, and so on. Eukaryotic cells, therefore, are more complex than prokaryotic ones, at least from a structural point of view. Fungi, protists, animals, and plants are made up of eukaryotic cells. Consequently, it is reasonable to

As early as the first centuries of life on Earth, this was characterised by exceptional biochemical complexity

The subsequent evolution of multicellular organisms proceeded mainly through increased structural complexity, whereas biochemical innovations were relatively few

assume that prokaryotic bacteria are ancestors of the eukaryotic unicellular organisms (such as numerous protists, including unicellular protozoans and algae). In their turn, the unicellular organisms are ancestors of multicellular organisms, that is, fungi, many protists, animals, and plants. In short, evolution progressed through a gradual increase of structural complexity. Consequently, turning points in the early development of life on the planet were the appearance of the first prokaryotic cells, the appearance of the first eukaryotic cells, and the appearance of the first multicellular organisms.

How do we travel to the past? The machine that goes back in time is supported by palaeontological records, rock dating, and the molecular clock

When did these events occur? Until recently, the only data available to help answer this question were fossils. These, combined with reliable methods developed to date fossil-containing rocks, led to some first-time estimates. The first data on the existence of microbial communities (the so-called stromatolites) indicate the events occurred 3.7 billion years ago, whereas the first traces of prokaryotic photosynthetic bacteria date back 3.3 billion years. However, the fossils of these fragile life forms lacking a hard skeleton are particularly scarce, and the further one looks in the past, the scarcer they become. Many of these fossils have been preserved so poorly that there are doubts as to their authenticity. However, the recent revolution in molecular biology has supplemented, so to speak, palaeontological findings with the invention of the molecular clock.

The molecular clock is based on the genetic distance between present-day organisms, and it is used to calculate the time distance from the moment of their deviation from a common ancestor

The molecular clock is based on the following principle: If the species deviated from a common ancestor at a given time, the evolutionary process of the speciation can be represented by a tree in which the common ancestor is the trunk and the descendants appear as successively diverging branches. Descendants evolve towards new species through accumulated mutations, alterations within the basic genetic code written in the DNA. Therefore, the new species emerging will be genetically differentiated, and the further in the past the species in question appeared, the greater is the genetic differentiation. Therefore, in comparing the genetic material of two groups, conclusions may be drawn not only about their possible relationships but also about the time that elapsed from the point of divergence of these groups from their common ancestor. By using palaeontologic records, that is, the morphologic similarity of fossilised organisms to their descendants surviving to date, assumptions can now be made as to possible relationships. Then, by analysing the degree of genetic differences from present-day organisms, one can draw possible phylogenetic trees and approximately date the appearance of particular branches. The conclusions of these recent studies with regard to major groups (kingdoms) of organisms changed previously held views on the two major kingdoms, prokaryotic and eukaryotic organisms. Briefly, the conclusions are as follows:

The so-called archaeobacteria, which live in extreme environments, were the first organisms on the planet

Prokaryotic organisms are divided into two major groups: archaeobacteria and eubacteria; genetic differences between them are as pronounced as those between prokaryotic and eukaryotic organisms. Archaeobacteria include the so-called extremophiles, bacteria adapted to living under extreme conditions of very high temperatures (up to 90 °C), exceptionally high salinity (up to the saltwater saturation point), and extreme acidity or alkalinity. These bacteria are found in hot springs, salt planes at the final stage of water evaporation and salt sedimentation, and, generally, in environments where other organisms cannot survive. Paradoxically, eukaryotic organisms, including plants, have greater phylogenetic affinity for archaeobacteria than for eubacteria. It seems even more paradoxic at first sight that photosynthetic bacteria, whether aerobic or anaerobic, belong to the kingdom Eubacteria, the group that is less related to plants, which photosynthesise. How can this be explained?

"Superior" organisms (plants and animals) are related more closely to extremophile archaeobacteria than to ordinary eubacteria

One of the problems with molecular analysis of phylogenetic similarities and differences concerns the molecule to be selected as the basis for such analysis. The conclusions mentioned earlier were based on the analysis of the so-called ribosomal RNA (r-RNA), a molecule somehow similar to DNA. DNA exists mainly in the nucleus of eukaryotic cells, but it is not absent from chloroplasts and mitochondria. Chloroplasts (and mitochondria) are semiautonomous organelles, that is, they include some, but not all, of the information and programming for their growth and function. The rest is contained in the DNA of the nucleus. Therefore, a chloroplast can survive outside a cell for only a very short time. For its proteins to be synthesised, a combination of messages originating from nuclear and chloroplast DNA is necessary. For example, the very important enzyme Rubisco, which is the catalyst for the photosynthetic assimilation of CO_2, is made up of two eight-part fragments: the first one is controlled by nuclear genes and the second by the chloroplast's own genes. The DNA of plant chloroplasts is identical to DNA fragments of surviving photosynthetic cyanobacteria, whereas the DNA of mitochondria is identical to fragments of surviving purple bacteria. Both bacteria types belong to Eubacteria, the group that is most remote from eukaryotic organisms.

However, some cellular organelles (chloroplasts, mitochondria, flagella) are more closely related to eubacteria

Why does this strange DNA identification exist? Why is it partial and not total? Chloroplasts and mitochondria have other similarities to bacteria. They also have a double external membrane and cyclical DNA (like a snake biting its tail), reproduce like bacteria, and are sensitive to bactericides. These similarities have been known for decades and were used to formulate a particularly fertile and challenging "endosymbiotic" hypothesis for the origin of eukaryotic cells. A strong confirmation of this hypothesis, which some scientists have elevated to a theory, came from the recent achievements of molecular biology, particularly the identification of chloroplastic and mitochondrial DNA with DNA fragments from photosynthesising cyanobacteria and purple bacteria, respectively.

The endosymbiotic episode or evolution by force of borrowing, usurping, or imprisoning genes

To simplify the endosymbiotic theory, imagine a heterotrophic prokaryotic organism permanently anxious about finding food in the form of probably hard-to-come-by molecules while a cyanobacterium in the same neighbourhood enjoyed the sun and assimilated, through photosynthesis, the ample atmospheric CO_2 into organic matter. The least aggressive solution would be to seek cooperation. More than likely, a little organic substance could be spared by this wonderful device that found a way to use solar energy and a simple, omnipresent molecule (CO_2) as a source of carbon. If the heterotrophic prokaryote offered to give the cyanobacterium something it lacked in exchange, then an interdependent, mutually beneficial relationship would emerge. Indeed, there are today, and there definitely existed in the past, communities of bacteria with different physiologies, and the close contact among them led to an exchange of services (mutualism). If an organism had nothing else to offer, the neighbour might allow it to use its waste, provided the organism did not bother the neighbour (commensalism). The organism might even become a "barnacle" (parasitism) and consume the neighbour, although this solution might not be in the organism's best interest, because it would have to use energy to break its resistance and then look for a new victim. As is true of relationships within human communities, among bacteria or other organisms there are all sorts of characters and many alternatives. The first case is an honest trade agreement (mutualism); the second is based on tolerance, charity, and kindness (commensalism); and the third is war, destruction, and looting (parasitism). However, what about hospitality?

Once upon a time in the history of life on Earth, a prokaryotic unicellular organism "swallowed" a whole prokaryotic cyanobacterium. The technical term for this is *endocytosis*, defined as the capacity of the cellular membrane to engulf materials from the environment and release them into the cell interior. The benefit of this hospitality is obvious: the host, that is, the previously heterotrophic organism, now has a guest with a light-collecting antenna that traps light and turns its energy into chemical energy. It also is equipped with enzymic catalysts that use this chemical energy to synthesise sugars and other complex organic compounds from simple CO_2 molecules. All these come in an accommodating, self-contained, ready-to-use packet in the form of a cyanobacterial cell. The formerly heterotrophic organism has now become autotrophic, not by constructing a photosynthetic mechanism from scratch, but by appropriating an already functioning one. Thus, the organism no longer needs to look for food or is limited to places where food is available, it can expand anywhere. The only requirement is light, that is, relative proximity to the water's surface. Carbon dioxide is everywhere.

Once upon a time, a heterotrophic bacterium hosted a photosynthetic bacterium. This way, it acquired a photosynthetic machine and a competitive advantage.

How would a host keep such a precious guest? What would the host do if the guest got bored and asked to leave? How might the host deprive the guest of any chance of escaping? What about keeping the guest as a prisoner? How would that be possible? Technically, the solution is called *horizontal gene transfer*, which in practice means that the host-cell removes some useful genes from the guest's DNA and transfers them to its own DNA. Therefore, the guest cannot escape and remains a handicapped prisoner forever because part of its genetic information is under occupation and ruled by the host, which now controls its sound operation and enjoys its products. This is why only part of the chloroplast DNA is identical to the DNA of free photosynthetic cyanobacteria. The rest of the genes are contained in the DNA of the host, which has become

Then the host imprisoned its guest by stealing its genes; in other words, the host robbed the guest of its capacity to survive independently and, therefore, to escape imprisonment

the prisoner's master. The prisoner's reproduction, full growth, and maintenance are controlled by the master. All the chloroplasts contained in plants and algae today were once free cyanobacteria originating from one or perhaps more episodes of successful imprisonment that occurred 2.3 billion to 2.1 billion years ago. Since then, prisoners have been passed down from generation to generation through the reproductive cells of their hosts. In angiosperms, for example, the female reproductive cells (ova) contain immature chloroplasts (called plastids), whereas the male reproductive cells contained in pollen grains have none. Plastids, therefore, are inherited from the mother and are then divided and distributed to the cells of the new plant; they mature into fully functional chloroplasts.

Most cellular organelles of eukaryotic organisms (i.e., protozoans, algae, fungi, plants, and animals) are the product of imprisonment and theft; they used to be free bacteria, and they are transferred to the offspring on the maternal side

Evidence indicates that the same thing happened with mitochondria, the organelles that perform aerobic respiration in the cells of all eukaryotic organisms, from protists to animals and plants. These once were free aerobic purple bacteria that were apprehended during about the same primal era and have been living in captivity ever since, giving their host the opportunity to acquire the aerobic metabolism package.

A third organelle believed to have been acquired in the same manner is the flagellum carried by several unicellular, eukaryotic organisms, as well as by specific motile cells of multicellular organisms, such as human spermatozoa. The flagellum is not a torture instrument but a motion organelle, propelling the cells that need to move fast. Flagella are similar to a category of Eubacteria called Spirochaetae. Similarities concern both morphology and structure of the tubular contracting proteins that induce movement. Spirochaetae move independently, propelling in a spiral motion. Theory has it that catching and keeping a spirochaete in captivity added one more organelle to the kit of eukaryotic cells, which evolved from prokaryotic ones with successive acquisitions of independent packets of genetic information.

The endosymbiotic hypothesis may explain the origin of chloroplasts, mitochondria, and flagella, but not the origin of organelles with a single membrane, such as microsomes and the endoplasmic reticulum, nor does it explain the origin of the nucleus, which has a double membrane and contains the chromosomes and the DNA.

However, because these events in essence are captivity episodes, why do they continue to be called endosymbiotic episodes? The reason probably is that scientists want to stress the huge positive impact they have had on the evolution of life on the earth, or perhaps because of the new system needed to transfer part of the control from the guest to the host to gain in stability. When a symbiosis is characterised by strong mutual benefit, the partners need to sacrifice only the tiniest part of their freedom, probably a negligible amount compared with the benefits gained. It cannot be easily claimed that the autotrophic cyanobacterium derived some benefit from the "hospitality" and therefore remained "on its own will," nor that a spirochaete that once carried only itself and now must propel its host as well has gained anything. How would a host entice a guest – who may have found its hospitality fun at first but then got bored and perhaps annoyed with the host's ulterior motives – into permanent cohabitation if it had nothing important to offer the guest? This is a difficult question. Another part of the book shows that true symbiosis in nature has three features: 1) there is a clear benefit for both partners; 2) the partners can also survive as individuals; and 3) the partners keep an eye on each other to ensure they are not short-changed.

To conclude the topic of endosymbiosis, it may be useful to mention that the result of this important evolutionary episode is that superior eukaryotic organisms have more than one genome. A genome is the sum of an organism's genes. Human genome mapping, which was completed in recent years, in essence concerns the genome of the nucleus. However, like animal cells,

We, therefore, have a multiple genome, with a diverse evolutionary origin

human cells have two sets of genomes, the nuclear one (which has main control) and the mitochondrial one. Besides the genomes of their nucleus and mitochondria, plants also have a third set, the chloroplasts. When symbiotic practices are discussed later in Chapter 8, it will be shown that certain organisms are even more complex with regard to the genomes that contribute to their health and survival.

Let us now return to that primal era when there were no plants but evolutionary phenomena, such as the endosymbiotic episode, shaped favourable conditions for further evolution of life forms, some of which eventually colonised the land as plants. Based on palaeontologic findings and the chemistry of rocks formed at that time, the sequence of events probably is as follows:

A short chronicle of life before the appearance of land plants

- Life appeared, and soon the first organisms were separated into Archaeobacteria and Eubacteria around 3.7 billion to 3.5 billion years ago
- Very soon, the first photosynthetic bacteria appeared, around 3.3 billion years ago
- Aerobic metabolism and the endosymbiotic episode are dated at 2.3 billion years ago
- The first eukaryotic organisms (the extinct gigantic unicellular alga *Grypania*) probably appeared 2.1 billion years ago
- The first multicellular algae appeared 1.2 billion years ago, whereas the class of multicellular Chlorophyceae (green algae), from which plants originate, appeared 800 million years ago. It took about another 400 million years before the first plants colonised the land.

Each of these milestones was accompanied by an increase in biodiversity, that is, the number of species encountered as fossils. The bursts of biodiversity seem to be associated with global and relatively abrupt changes in atmospheric O_2 and CO_2 and temperature fluctuations, as described elsewhere in this book. The long intervening periods were relatively calm and stable with regard to the number of species. What is impressive, though, is the short period of about 200 million years from when the waters of the first oceans acquired mild physical–chemical conditions compatible with life to when the first microbial organisms appeared. However simple the form of an archaeobacterium may be, its biochemical makeup is as complex as that of a human cell. How was it possible for all metabolic pathways – inherited later by organisms descended from these pioneers – to emerge so fast and be tested by natural selection? How was it possible, within this relatively short time, for structures and metabolic pathways to be organised into cells with complete growth patterns, into cells that divide accurately so as to transfer to their descendants the necessary genetic information, into cells that perceive environmental changes and respond to them? The difficulty in finding a convincing answer to these questions led to the formulation of the panspermia hypothesis, which claims that life on Earth was imported from somewhere else in the universe. The candidate carriers proposed are exceptionally hardy bacterial spores travelling inside meteorites or comets, thus protected from intense ultraviolet and cosmic radiation, the deep cold of interstellar space, and the high temperature of the collision's impact. Any spore that survived would have at its exclusive disposal a whole virgin planet. This is a charming and provocative view that simply shifts the problem of life's origin beyond the earth. Whether indigenous or imported, this original simple life form evolved into the incredible diversity of organisms observed today in nature or at museums, which, if humans are careful, will continue to evolve into the future.

Outbursts of biodiversity and global climatic changes

Back to the dawn of time: and the origin of life is?

How was it possible, in the primeval ocean, for basic cellular structures and functions to become established and stabilised so soon?

Perhaps the view that life was imported to earth is not a fairy tale. Of course, it did not happen through strange extraterrestrial humanoids; some bacterial spores inside meteorites may have been enough.

Required Structural and Functional Innovations that Enabled Land Colonisation

How was the land colonised?

What were the necessary prerequisite conditions? Soil formation, atmospheric oxygen, and the ozone shield

The land was colonised 440 million years ago

The first land plants originated from freshwater algae

The next big step to be described is land colonisation. The prerequisites for this event were suitable and sizeable littoral areas with some weathering – however limited – of rocky coasts so that a little soil could be formed (the plants would be responsible for the rest), as well as climatic and atmospheric conditions compatible with life on land. Readers are referred to Chapter 3 for details about the prevailing atmospheric conditions at that time. A reminder might be appropriate here that the photosynthetic activity of aerobic cyanobacteria and algae had already increased the atmospheric concentration of O_2 to acceptable levels (around 10 %). Therefore, the ozone layer must have been created in the stratosphere as a result of the effect of solar ultraviolet radiation on O_2. The ozone layer was the necessary shield that would protect the first land colonists from harmful ultraviolet radiation. Furthermore, because of the photosynthetic activity of bacteria and algae, as well as geologic factors, atmospheric CO_2 was adequately reduced so the greenhouse effect would not be as intense as during the first stages of the earth's history. The oxygen increase had already favoured the much more efficient aerobic metabolism, allowing the appearance of complex animal species and algae in the oceans. At the end of the Ordovician period (around 440 million years ago), the need for the creation of littoral soil likely had been fulfilled by the combined action of acid rain and acid secretions from microbes and lichens that caused rock detrition. The climate was cold, because it was the end of an ice age, and quite humid. This is the period when the first land plants appeared; it is believed they originated from multicellular freshwater green algae.

The passage from water to land was a huge step requiring significant changes and innovation, both in plant form and in plant function. Two important problems had to be resolved: the risk of dehydration and the effect of gravity. To perceive the scale of these problems and appreciate the solutions selected, one must compare the corresponding environments and lifestyles of algae and their plant descendants. This analysis is limited to the so-called benthic algae, those attached to some substrate, only because they are relatively similar morphologically to plants. Recall that there are multicellular and unicellular algae that are not attached but that float in water. Both algae and plants are autotrophic photosynthetic organisms that use light as an energy source and CO_2 as a carbon source and that also need certain mineral nutrients and water. Plants take mineral nutrients and water from the soil, meaning they need roots, which also anchor plants to their position. Algae are surrounded by water containing dissolved inorganic elements that can be absorbed by the whole surface of the algae. Therefore, algae do not need roots; the base of their thallus (the technical term for the body of the algae) simply forms an organ that attaches the algae to the solid substratum and ensures they do not drift with the currents.

A great leap to an inhospitable world: dehydration and gravity

A comparison of land plants with their closest aquatic ancestors, or how the new environment made it imperative to develop innovations

Roots: an answer to the problem of support and absorption of localised nutrients

Algae are not at risk for dehydration in their natural environment. However, they suffer from it as soon as they are exposed to the free atmosphere. To respond to the dehydration threat, plants developed the so-called cuticle, a thin, waxy, extracellular layer that complements the plant's epidermis and is impermeable to water. Algae have no cuticle, and their epidermal cells are directly exposed to water so they can receive the dissolved CO_2 and mineral nutrients. From a chemical point of view, the cuticle is made up of complex polymers of fatty acids with the generic name *cutin*, meaning the biochemical equipment (e.g., enzymes) necessary for cutin production had to be developed.

Wax on the epidermis prevents dehydration

Fatty acids are produced inside the cell, but the cuticle is an extracellular formation. Therefore, methods to control fatty acid exportation had to be invented so fatty acids could be delivered to the right sites in the right quantity to ensure construction of the cuticle.

Another problem is offspring dispersal. Within water, algae spores are not at risk of dehydration. On land, this is a significant hazard, even though the first land colonisers lived in the littoral zone and spores landed on wet soil. It would not have been possible for early plants to move away from this zone before adapting to life on land and achieving spore protection. In that sense, the possibility of producing water-impermeable cutin (cuticle), which envelops not only the whole of the plant body but also its spores, was particularly important.

However, epidermal impermeability limits the entrance of CO_2

Although the cuticle provides adequate protection against dehydration by blocking the exit of water vapour, it also prevents the entrance of CO_2, which is essential for photosynthesis. The cuticle, therefore, solved one problem but created another. The solution finally selected was a compromise between the need for photosynthesis and the need to prevent dehydration, meaning a cuticle interrupted by pores through which atmospheric CO_2 can enter the plant. These pores are called stomata and are found mainly in photosynthesising organs. However, water vapour exits through the stomata, which means the risk of dehydration remains. Suppose one wanted to refresh the air inside his or her home without making the house unacceptably cold. He or she could set an acceptable temperature level and an acceptable level of air renewal, that is, a compromise that would allow the windows to be left open more or less. One also might open the windows only when the external temperature is at its highest, for example, in the early afternoon. In other words, one would adjust the opening depending on a compromise between the two opposing needs.

Stomata: valves that reconcile two opposing needs

This is exactly the mechanism plants adopted. The stomata are flanked by two specialized epidermal cells that can change shape and thus regulate the size of the opening from wide to tightly shut, including all stages in between. As is shown in Chapter 6, stomata cells are equipped with sensors that "measure" the quantity of light, CO_2, and water vapour. In this way, they perceive whether it is night or day, it is cloudy or sunny, or the atmosphere is dry or humid. Furthermore, the stomata receive signals from the root, informing them about soil water reserves. All this information is processed so they may open to the size most appropriate for the conditions prevailing at any given time. The aim is to achieve the maximum photosynthesis possible with the least amount of water loss. In other words, the entire control system is based on compromise, at least when environmental conditions are not extreme. However, when survival is at stake, such as during extreme drought conditions, the stomata lock. The stomata also remain closed throughout the night. Given that in the absence of light there is no photosynthesis, there is no point in losing water. Such mechanisms of regulated valves are not necessary for algae. Therefore, algae have no stomata.

The stomata are sensory organs that measure environmental parameters, process information, and appropriately regulate their openings. If conditions are suitable, they open and the plant photosynthesises. If not, they close so as to limit the risk of dehydration.

A third innovation was necessary to cope with gravity – that is, the absence of buoyancy. Using a diving mask, one can see benthic algae standing upright, swaying under the effect of currents or waves. If one of these algae were sacrificed and taken to land, it would not be able to stand upright and would collapse under its own weight. Therefore, the development of supporting systems was absolutely necessary; so cellular walls were reinforced and became thicker, whereas a series of hard tissues called *sclerenchyma* strengthened the stems. Sclerenchymatous cells become tough through the deposition of lignin on their cellular walls. Lignin is a phenol polymer, stemming from the phenylpropanoid pathway. This is a biosynthetic system

Thick cellular walls are an initial response to the force of gravity

with great potential for producing a multitude of organic substances based on the phenol ring. Such substances play an important role not only in plant support but also in plant defence and ecosystem balance (see more in Chapter 7).

These initial adaptations to life on land seem to have developed gradually over a period of about 20 million years, starting at the end of the Ordovician period (440 million years ago). It is assumed that adaptation processes took place in littoral freshwater areas because in such regions there are frequent fluctuations of water masses. Water is reduced during intense drought and high-temperature periods, as a result of evaporation, whereas it increases with the addition of water masses due to rain or water flowing from adjacent areas of higher altitudes, when the climate becomes more humid. The water level rises and falls, which means that benthic algae close to the coastline are alternately covered and uncovered. One may reasonably assume that if a series of mutations appeared in some of them and a feature that would enhance their tolerance to dehydration emerged, such algae would have an advantage over others. The new features would become permanent because of natural selection. This new, enhanced genome and the organisms carrying it would be capable of leaving the coastline somewhat and moving towards the hinterland, probably to exploit new opportunities.

Was it worth the trouble, though? Did land life offer advantages?

Were there any new opportunities? Was there any advantage in the new lifestyle allowing the algae to leave the security of water? Some simple physics help show it was well worth the effort.

Imagine a multicellular alga that photosynthesises underwater. It receives the CO_2 dissolved in the water through its entire body. Every CO_2 molecule it receives from its immediate environment leaves a void that needs to be filled by a new CO_2 molecule so that photosynthesis may continue. This new molecule will come from some distance and will take some time to reach the surface of the alga. Every photosynthesising alga, therefore, creates a local reduction in CO_2 concentration in its immediate surroundings – in essence, a graded concentration that increases the further away from the alga surface. The driving force leading to CO_2 concentration gradients is photosynthesis, and the force that tends to restore a similar concentration by transporting CO_2 to the alga from a distance is diffusion. The density of the medium is a parameter of resistance against diffusion. In simpler terms, a CO_2 molecule tends to move spontaneously from a high-concentration region to areas of lower concentration. Along the way, it collides with the water molecules in between, causing delays. If there is air instead of water between the two sites, then the movement will be faster because the collision probability rates are significantly reduced. In numeric data, the diffusion velocity of CO_2 in the air is 10,000 times higher than in water. This means that photosynthesis of a plant submerged in water is limited by the slow diffusion of CO_2, whereas such diffusion in the air is instantaneous. Hence, when land living was adopted, the photosynthetic limitation previously imposed by the density of the water medium was removed.

The problem of gas diffusion

Photosynthesis is faster in air than in water…

A second benefit involves light. Although a glass of water seems totally transparent, a long column of water absorbs some light. The phenomenon increases because of particles floating in the water or organisms absorbing or dispersing light. A glass of fresh water from a lake with a relatively high plankton density is not completely transparent. Land colonisation, therefore, allowed photosynthetic organisms to enjoy more light and increase their photosynthetic capacity.

…besides, there is more light

Based on this scenario, the first land plants were similar to the green algae from which they originated, but they also were characterised by certain adaptations that made them more resistant to dehydration. Undoubtedly, the initial forms were fragile: small plants not yet fully liberated from the need to be close to water. Size increases entailed more innovation. Imagine a crowd of small plants a few centimeters tall, engaged in a biological war because of competition for more light, trying to avoid one another's shade. Growing upwards, however, might offer a competitive advantage with regard to light collection but would distance the photocollecting tissues from the wet soil. Water and minerals a tiny crawling plant may absorb directly from the wet soil solution are now further away from the plant parts that moved higher to photosynthesise better.

Inevitably, the first land plants were small and fragile

Competition for light and the need to grow in height

The solution that characterises all organisms, when their size increases, is specialisation, the creation of organs that undertake different functions and a system of communication among these organs. This is how the root developed, as an organ to absorb water and nutrients, which, however, had to be forwarded to aboveground organs at higher risk of dehydration. This led to the development of vessels (the xylem vessels) transporting water and mineral nutrients, and the solution selected was one that also served as a support system. These vessels have ligneous walls that keep plants upright. The system also was enriched by complementary hard tissues capable of providing additional mechanical strength to the stem. Finally, the annual development of new ligneous vessels while the old ones stopped functioning but still existed (see the discussion on annual rings in Chapter 3) enabled relatively tall woody plants to appear gradually. However, the root and trunk are heterotrophic parts of the plant; they cannot photosynthesise and depend on the products of photosynthetic organs. Therefore, peripheral to the xylem vessels, in the so-called bark of the trunks or stems, there are other vessels (the so-called phloem) that transport useful organic substances,

Innovations that contributed towards a height increase

Vessels as tissues for the communication between distant organs

Woody tissue: an invention for support that made it possible for leafage to rise up towards the light

rich in energy, to the roots. In other words, there is an allocation of tasks, which is readily observed in every multicellular organism with increased size.

The last necessary innovation concerns reproduction, which also must adapt to drought conditions. In algae, the male reproductive cells are released by corresponding male organs and swim to the female ones (ova) – usually of another individual of the same species – with the help of a flagellum. When the male cells find the female ones, the two cells fuse and a new organism is created. The reader might wonder how they manage to get to the right place, how they are not misled into approaching the ova of another alga species. Once humans find their mate, they know exactly what to do. How do plants manage? Even if a mistake is made – also known to happen among humans – why are no monsters, chimaeras, mermaids, or centaurs born? These questions are explored later, in the discussion of the sexual life of plants.

Reproductive innovations

Land environments are fraught with dehydration hazards for motile male reproductive cells. Some land plants have not resolved this issue fully, so they can have sex only when it rains. These plants are the bryophytes (mosses) and pteridophytes (ferns). Many pteridophyte species live in dry environments, yet reproduce only when it rains. Gymnosperms (conifers) and angiosperms (flowering plants) found a solution: they envelop male reproductive cells inside a tough outer casing. Indeed, one of the substances participating in the structure of this tough wall is the sporopollenin of pollen grains and spores. This substance is the champion of resistance to degradation among all substances ever produced by living organisms; it may remain intact for hundreds of millions of years. It is so resistant to degradation that pollen grain walls remain intact even if the rock that contains them is artificially shattered with potent acids. The morphology of pollen grain walls is characteristic of the plant species that produced them.

Sporopollenin: a biomolecule with exceptional resistance to degradation

Primal pollen grains, or how we know the composition of past ecosystems

Therefore, on the basis of primordial pollen grains, a picture of past vegetation in various parts of the earth can be reconstructed, even if there are no other plant fossils. If DNA were as resistant as sporopollenin to degradation, the idea behind the movie *Jurassic Park* – that is, reviving extinct life forms – would have some likelihood of success.

Today, in this era of mass extinction of organisms due to human intervention, the ability to revive them in the future might be achieved by creating so-called gene banks, through the controlled freezing of reproductive forms such as spores and seeds. At this point, the following questions may be raised: Have there been mass extinctions of species in the past? If so, why? If they happened, when did this occur? If it occurred, did extinction of animal and plant species coincide?

Interactive Systems: Plants, Climate, Animals, Soil

Plants were the pioneers that paved the way, leading to land opportunities for the animals

Within a mere 50 million years, the earth was covered by forests of giant pteridophytes (ferns)

Mass extinctions are probably the third most critical episode in the evolution of life on earth, following the endosymbiotic incident, which allowed the creation of eukaryotic cells, and land colonisation by plants. This colonisation meant that the land was filled with ready-made plant food, making it possible for animals to ensue. The basic plant innovations were concluded, enriched, and perfected within a few million years, with plants spreading in every land surface available; this in turn meant a modification in the climate and the water cycle, a reduction in CO_2 and an increase in O_2 in the atmosphere to tolerable levels, and significant regulation of the earth's temperature through plant activity. In less than 50 million years, the earth was covered by forests, mainly of giant pteridophytes (ferns). The first angiosperms (flowering plants) appeared rather late, about 140 million years ago, or almost 300 million years after land was colonised. Before angiosperms appeared, the earth, at least as far as plants were concerned, must have been quite boring in colour because there were no flowers. The appearance of flowering plants, the increase in the number of their species, and, finally, their prevalence occurred quickly, within a few million years. Why did it take so long for them to appear? Three hundred million years of gymnosperm and pteridophyte prevalence is a very long time. Two reasons have been proposed. First, the appearance of flowering plants (the first of which were small herbs) seems to have coincided with a change in the type of herbivorous dinosaurs that dominated the earth. Mammals had not appeared yet. At that time, dinosaurs changed their dietary habits somewhat. Until then, the herbivorous dinosaurs stood on their hind legs and fed on the foliage of tall trees. Therefore, adult conifers and giant pteridophytes were under high grazing pressure, whereas younger ones were not affected. This meant there was no interference with forest renewal. However, around the end of the Jurassic era (i.e., 144 million years ago), a high concentration of dinosaurs standing on their four legs appeared, meaning they fed mainly on young conifers, which reduced the conifer populations and cleared some space for the development of primitive flowering plants. The latter may have been food for these dinosaurs as well. However, because these plants were small, they had a shorter life cycle; therefore, a small but reproductively mature flowering plant was quite likely to produce offspring before being eaten. In contrast, the slow-growing conifer was less likely to do so. Consequently, a change in the dietary habits of herbivorous dinosaurs tipped the competition between the evolutionary older conifers and the newcomer flowering plants in favour of

Flowering plants appeared much later, 140 million years ago; their flowers broke the green monotony

Why is the appearance of flowering plants associated with the appearance of dinosaurs walking on all four legs?

Coevolution of insects and flowering plants

the latter. A second hypothesis associates the appearance and dispersal of flowering plants with the appearance of insects involved in their pollination. This is how the presence of flowers favoured particular types of insects and these insects, in turn, favoured the spread of flowering plants. This is a case of *coevolution*.

Another group of plants that appeared relatively late is that of grasses (Poaceae). Their presence on Earth played a huge role in the evolution of humans. Today, this group includes around 10,000 species. Among them are the ancestors of the first plants tamed by humans – wheat and barley – which were domesticated around 11,000 years ago in the region of southwest Asia where Iraq, Syria, and Iran are located today. Others include rice and millet in China; sugar cane in New Guinea, 9,000 ago; sorghum in sub-Saharan Africa, 7,000 years ago; and maize in Central America, 5,000 years ago. The taming of Poaceae was one of the reasons the first permanent human settlements appeared, because Poaceae cultivation not only allowed permanent human residence, it made it imperative. Along with the taming of goats and sheep in the same region (Iraq, Syria, Iran) during the same period, Poaceae cultivation made it possible, for the first time in their history, for humans to store food for difficult times and allowed them to give up the nomadic life of hunter–gatherers.

Poaceae (grasses) appeared after the extinction of the dinosaurs, 65 million years ago

Poaceae belong to the flowering plants, which, as stated earlier, appeared 140 million years ago. The first traces of Poaceae date to the beginning of the Eocene period (60 million to 55 million years ago), just after the worldwide hecatomb that caused the extinction of most animal species, including the dinosaurs, 65 million years ago. It was then that mammals made their first timid appearance, free from the dominance of the dinosaurs. However, the population densities of the first Poaceae were very low and the number of species very few. These plants developed much later, during the

Miocene period – 10 million to 20 million years ago – when the first extensive low grass vegetation ecosystems, in which Poaceae prevailed, started appearing. What favoured their extensive spread in the past 10 million years? An interesting view associates their spread with the prevailing drier climate and the increase in populations of herbivorous mammals grazing on low vegetation. Indeed, the Poaceae tolerate drought. Furthermore, their leaves grow in an idiosyncratic manner: there is interstitial growth at their leaf base. At this base, the meristematic cells keep dividing so that leaf blades increase continuously. However, this is not true for the rest of the plant species, in which leaf growth is homogenous, involves the whole leaf surface, and stops when the leaf reaches its final size. In many Poaceae species, the leaf's base is quite close to the ground or even within the soil. (This does not apply to the familiar cultivated Poaceae, which differ significantly from their wild ancestors.)

However, they were established in extensive populations just 15 million years ago. What took them so long?

Consequently, the top of a leaf may be eaten for the most part or burnt by fire, but the leaf does not stop growing from its base. This is the reason the lawn can be mowed and grow back fresh a few days later. The drought of that era often led to spontaneous fires, and herbivorous mammals applied a great deal of grazing pressure. However, the Poaceae recovered better from both pressures compared with the other plants. Therefore, they evolved significantly and increased their biodiversity and species numbers. Among these species appeared those with properties making them more suitable for domestication; these species now comprise 50 % of the food currently consumed by humans.

Droughts, fires, grazing pressure, and a noteworthy leaf-growing pattern among Poaceae

Violent Environmental Perturbations and their Evolutionary Significance

Mass (global) extinctions of species as evolutionary phenomena

Regarding the question of mass catastrophes, what is meant by this term? *Mass catastrophes* are defined as episodes of fast extinction of species (in a few million or a hundred thousand years); these episodes are global and involve a large number of species, and the extinct species are distributed in many higher-classification units, such as families. In this sense, an episode involving only birds is not considered a mass catastrophe, even if it is global. Nor is a mass catastrophe an episode that affects all major animal groups in only a small region. It also should be noted that recognised mass catastrophes have been based on animal palaeontologic records and comprise the following five periods:

When did they occur, and what were their consequences?

- 443 million years ago, when 96 % of marine animal species disappeared
- 364 million years ago, when 80 % of marine animal species disappeared
- 248 million years ago, when 90 % of marine animal species disappeared along with 70 % of the families of land vertebrates
- 206 million years ago, when 40 % of marine animal organisms and a large percentage of insects disappeared
- 65 million years ago, when 80 % of marine animal organisms disappeared, with a drastic reduction in land mammals and the total extinction of dinosaurs

What were the reasons – terrestrial or extraterrestrial?

What are the reasons behind mass catastrophes? Obviously, there must have been exceptionally violent events, because they affected the whole planet and all organisms regardless of evolutionary distances, from tiny marine crustaceans to gigantic dinosaurs. Many hypotheses have been put forward pointing to terrestrial or even extraterrestrial causes. The most probable among the terrestrial causes involve a reshuffling of the earth's crust accompanied by shifts of continents, mass global volcanic activity, and significant changes in sea level. If, for example, a continent moves towards the poles or the equator, the climate will become colder or warmer, respectively. Mass volcanic explosions, beyond local disasters, lead to abrupt cooling of the planet's climate because the penetration of solar radiation is obstructed by the clouds of dust and gases spreading throughout the atmosphere. As a result of the condensation of atmospheric water vapour, after the temperature drops, there are strong acid rains because the raindrops dissolve volcanic gases; subsequently, ocean waters become acidic. Furthermore, concentrations of toxic chemical elements increase on the earth's surface following the precipitation of flying ash. A long volcanic winter is followed by a greenhouse effect and planetary warming because of the increased CO_2 quantities in the atmosphere. Volcanic explosions of this scale, which affect the climate of the entire planet for several thousand years, have not been recorded in historic years. Geologists, however, have documents showing that at least twice in the past, the reshuffling of continents resulted in extensive temporal and spatial mass volcanic explosions and accompanying climatic changes. The formation of the gigantic supercontinent Pangea 250 million years ago was accompanied by mass volcanic explosions over an extensive area where present-day Siberia lies. Similar phenomena seem to have occurred around 65 million years ago, as indicated by the huge accumulation of basaltic lava in the area of the Deccan Traps in India. The third and fifth mass extinction episodes, therefore, might well have been the result of long-term, extensive volcanic activity.

Continental shifts, mass volcanic explosions, and changes in sea level

Consequent climatic changes, volcanic winters, and past greenhouse effects

A second group of mass extinction causes are those of extraterrestrial origin. At various times, there have been hypotheses involving comet storms, radiation

A meteorite terminated the long dominance (for almost 200 million years) of dinosaurs and allowed the development of mammals

from supernova explosions, solar spot intensification, and asteroids colliding with the earth. The most popular cause, which is supported by evidence, is a collision with an asteroid. Rock beds, dated to around 65 million years ago (i.e., when the last major species extinction – which included the dinosaurs – took place) contain high percentages of iridium, an element that is rare on the earth but ample in asteroids. These rocks are located in various parts of the planet. Such dating has been performed at the huge – 10 km in diameter – subterranean crater of the Yucatan in Mexico. It is believed that 65 million years ago, an asteroid 10 km in diameter hit the earth at a velocity of 40,000 km per hour. Such a collision releases huge amounts of energy, causing direct, indirect, and long-term effects on organisms. Direct causes of species extinction are the explosion and heat waves, the tsunamis, and the intense fires. The ash from the explosion rises to the atmosphere and remains there for months or years, resulting in a harsh winter because the sun's rays cannot reach the earth's surface. The ultra-acid rain transfers toxic elements from the flying ash to the ground surface and makes the oceans acidic. After the initial tsunamis, huge masses of water are trapped in the hinterland, causing extensive, long-term floods. Finally, when these phenomena ebb, a long-term greenhouse effect and planet warming ensue because of the CO_2 released into the atmosphere from the incineration of huge quantities of organic matter.

Why was the word plant not mentioned in the last few pages?

If one rereads the preceding description of the history of violent catastrophes that hit the earth during the last 450 million years, he or she will notice that the word *plant* does not appear; indeed, this history concerns extinctions of animal species. Even the well-known geologic aeons were defined, to a large extent, on the basis of mass extinctions of animal species. In chronologic order, the five major extinction events define, respectively, the transits from the Ordovician to the Silurian, the Frasnian to the Famennian (within the late Devonian), the Permian to the Triassic, the Triassic to the Jurassic, and the Cretaceous to the Tertiary. Every violent episode meant the destruction of previous ecosystems, the disappearance of previous species, and the creation of new habitats and opportunities for further evolution of the organisms that survived. Following the major disaster of 248 million years ago, the biological and ecologic void was filled by reptiles, the terrible lizards or dinosaurs, which survived one catastrophe in the meantime (at 206 million years ago) before they became totally extinct 65 million years ago. The new void was filled by mammals, which increased in size, number of individuals, and number of species and have reached the present era with the indisputable domination of humans. Mammals coexisted with the dinosaurs for a long time; however, they were small and insignificant, unable to compete with the giants with which they shared the planet. They remained in ecologic obscurity until the asteroid presented them with their opportunity. If global catastrophes are inevitable (which is implied by the fact that they happen every 50 million to 100 million years), then the master of the future earth should be sought among the organisms less favoured and less prominent today. The following questions hypothetically might have been raised by an intelligent being during past global disasters: Who survived, relatively unaltered, all previous disasters? What are the most obscure and disregarded living beings today? How did plants react to previous disasters?

The known geologic aeons of the earth's history were devised on the basis of mass extinctions of animals

Which organisms will survive the next major catastrophe?

Palaeobotany (the study of plant fossils), on the basis of its own criteria, also has divided the earth's history into three aeons. The first is that of the palaeophytes (400 million to 240 million years ago), which started with land colonisation. During that aeon, the prevalent species were plants similar to bryophytes; pteridophytes (ferns), which grew to tree-like sizes; and the

How could the history of the earth be divided in aeons if plants were more popular than dinosaurs?

lycopodiophytes. That aeon ended with the appearance of conifers, Ginkgophyta, and Cycadophyta. During the second aeon, that of the mesophytes (240 million to 100 million years ago), these three groups prevailed, until, towards its end, they lost ground with the appearance of the angiosperms (flowering plants). The last aeon, that of the cenophytes, is characterised by the prevalence of flowering plants.

Based on the history of plants, the dividing lines between geologic aeons concern changes in the prevalence of various plant groups rather than mass extinction episodes

Two basic differences should be noted with regard to the way the two lines of research – those of palaeozoologists and those of palaeobotanists – divide geologic aeons. The former focus on extinctions, while the latter hardly mention extinctions at all. Instead, they use as reference points the "prevalence" of certain classifying units as opposed to others, which, however, do not become extinct. In the palaeozoologic records, there are many cases in which large groups of animals completely disappeared – the most popular is the case of the dinosaurs. Those familiar with palaeontology also might mention other big animal groups that once were dominant but later disappeared, leaving no descendants in a blind-alley evolutionary line. Trilobites and ammonites are two such examples. In contrast, almost all large plant groups that prevailed in the past still have living descendants; in some cases, there are plant genera or species that have remained unaltered – at least from a morphologic point of view – for hundreds of millions of years. Remember how the discussion about the history of plants started at the end of the last chapter? It was with a reference to these "living fossils." Remember the *Ginkgo biloba*, a tree whose appearance has remained unchanged for 240 million years? It is only one of the tens of plant species included in the living fossils, illustrating the effectiveness of plant survival strategies. The corresponding living animal fossils are far fewer and only similar to (i.e., not exactly the same as) their fossilised ancestors.

A second difference between palaeozoologists and palaeobotanists in the way they divide the geologic aeons has to do with time. In other words, the transition times from one era to the next do not coincide, as if the evolution of plants and that of animals do not go hand in hand. Therefore, another major question is raised: How did plants react to the major catastrophes of the past? Were there mass plant extinction incidents? Did they coincide with the disappearance of animal species?

Geologic aeons based on the history of animals do not coincide with those based on plant history

Why did Plants not Suffer Mass Extinction?

A careful study of palaeobotanic records indicates that during major global disasters, vegetation lost significant ground. This is reasonable, considering the violent character of the phenomena. For example, evidence exists indicating there were huge, almost global, fires after the collision of the asteroid that brought about the end of the dinosaurs. Soon, however, the vegetation recovered and its composition generally was the same as it was before the catastrophe. In other words, there have not been significant incidents of extinction among plant species – they disappeared only temporarily (i.e., they are not found in fossil form) and reappeared later in the same territories. Although some species probably give way to benefit others, extinctions are negligible, are not global, and do not involve superior classification units. The same thing happened during the previous two global catastrophes. The genus *Equisetum* (plant) survived three disasters; the species *Botrycoccus braunii* (green alga) survived all five. Around 70 plant species have survived at the genus level for 100 million to 300 million years.

During the episodes of mass animal extinction, vegetation lost ground but recovered relatively soon and acquired more or less the same composition it once had

The answer, therefore, is clear: plants have not suffered mass extinctions, only temporary loss of ground. This leads to another question: What is the reason plants

Champions of survival and stability

do not suffer mass extinction? What is it that helps plants recover after their initial shock? What are the properties that equip them with this special survival capacity?

Why have plants not suffered mass extinction, at least not to an extent similar to that of animals?

Before this question is answered, and to avoid any misunderstanding, it must be stated that there have been, there are, and there will be extinctions of plant species. Readers more familiar with the subject know that many of today's botanists dedicate themselves to lovingly recording threatened plant species in the "red bibles" of species at risk for extinction. The question is mainly quantitative. The difference in biology between animals and plants indicates that animals are more susceptible to a disruption in their environment and, therefore, are more likely to suffer extinction. Indeed, this is what happens, as documented not only by the palaeontologic records but also from studies of animal and plant extinction rates in modern times. A truly intelligent argument to illustrate this point is as follows: If you bombard a zoo and a botanical garden today, the botanical garden will recover to more or less its previous state in a few years, but the zoo will never recover.

What makes plants so hardy? Let us return to the basic differences between animals and plants described in the beginning of this book: a world without animals is realistic because plants are self-sufficient and their nutritional requirements – light, inorganic minerals and water from the soil, and CO_2 from the atmosphere – are amply supplied by nature. With these simple, cheap, and omnipresent materials, plants grow and develop; they need nothing else. In contrast, animals are heterotrophic; they suffer from the lack of organic nutrients produced by plants, which are the primary producers of the biosphere. A world without plants is simply impossible, at least in the diversity of forms and the size of organisms to which we have become accustomed. The first life forms on Earth might well have been heterotrophic bacteria, but they must have been limited to a few

Returning once again to the basics of plant biology

spots where high-energy inorganic molecules locally emerged at the bottom of the oceans from cracks in the earth's crust. The appearance of photosynthesis allowed life to expand to every possible part of the planet, through the use of an extraterrestrial and inexhaustible energy source (the sun) and the exploitation of the most ample and simple materials, water and CO_2. Adding to these, the production of O_2 by plants results in the possibility of the creation of a cosmos.

Plants are autotrophic, that is, nutritionally self-sufficient

Following a disaster, therefore, plant and animal populations are decimated. Of the few remaining, the major problem of finding food obviously will be faced by heterotrophic organisms, populations of which will dwindle further because of food scarcity. Animal ecologists are very familiar with what is called the *minimum viable population*, the minimum number of individuals of a species within a territory below which the population cannot survive. Among animals, this number is quite high; among plants, however, the minimum viable population may be very low, particularly among trees and bushes, which grow to very old ages. Hence, one should keep in mind that famine following a catastrophe is very common among animals but particularly rare among plants, and that a reduction in population numbers leads to the extinction of an animal species much more readily than it does in a plant species.

The minimum survival population for plants is much smaller than that of animals

Another attribute related to plants' greater survival capacity is phenotypic plasticity. As explained at the beginning of this book, the size and shape of plants are not stable. As soon as they reach maturity, humans and all other motile animals reach their final size, with some small deviations from mean values. Furthermore, their bodies present a bilateral external symmetry, except for a few that present higher-class symmetries, such as the five-sided (pentaradial) symmetry of echinoderms (starfish and sea urchins). Plants are not symmetric, and they grow in accordance with the local availability of environmental resources or the local intensity of

Plants, as nonmotile organisms, have great phenotypic plasticity

environmental pressures. The plant part aboveground will turn towards the light, whereas the underground part will grow towards the most ample supply of water and mineral salt nutrients, even if this growth might follow a tortuous course. Where the winds are of a prevailing direction (e.g., to the south), the plant will turn in that direction. Technically, *phenotypic plasticity* is defined as the ability of a genotype (i.e., the total number of genes determining an organism's characteristics and growth pattern) to produce various phenotypes (i.e., forms), each of which is suited to a particular environment. In a sense, the genotype is an organism's "being" whereas the phenotype is its "appearance." In a plant species, for example, the underground part might weigh twice as much as the aboveground part when the plant grows in fertile and humid soil. If, however, conditions change and soil nutrients are exhausted, while there is also a drought, the plant modifies its form by increasing the underground part and limiting the aboveground one, perhaps resulting in the former being ten times the size of the latter. The same is true of other plant organs. If half a plant is in the light while the other half is in the shade, the leaves on the shady side will be larger and greener to increase their light-collecting capacity. These leaves also will be thinner, because the little light they enjoy will be absorbed by the top layer of cells, leaving none for any deeper layers. Therefore, there is no point in developing a multilayered leaf because deeper cellular layers would be redundant. This possibility for a plant to modify its growth according to current or local conditions is true not only of its morphology, but also its chemistry and physiology. All these are expressions of the incredible phenotypic plasticity of plants, of their developmental flexibility. In essence, plants can modify their inherent genetic programming depending on the resources available and the environmental pressures they are facing.

Some typical examples

"Being" (genotype) and "appearance" (phenotype) among plants

Phenotypic plasticity and exploitation of resources

In contrast, the genetic programming of animals is more rigid. A child who is not fed properly may become a small-sized adult, a few centimeters shorter than average, and may be susceptible to diseases. If the child's diet were restricted further, he or she might not even reach adulthood. Under the same circumstances, a plant will grow to the extent available resources allow, and only in extreme conditions will its health suffer. Increased phenotypic plasticity is a response to the lack of motility. Committed to leading its life fixed at one spot, the plant will spread as far as it can. It will become bolder if the environment allows, or it will limit itself to a humble existence under pressure. It will change its habits if the environment changes. It is not a coincidence that animals that, like plants, have adopted a sedentary life (e.g., sponges and corals), also present enhanced phenotypic plasticity. Therefore, the range of phenotypic plasticity reflects the corresponding range of conditions under which an organism may survive. Organisms with wide growth and developmental flexibility tolerate environmental changes more easily, even if these are extreme.

Developmental versatility and modification of one's inherent growth pattern: a comparison between plants and animals

Living humbly is an advantage

The previous example showed that a plant that is half in the light and half in the shade has two types of leaves, each suited to its circumstances. If the whole plant suddenly finds itself in the light, how would it react? The leaves formerly in the shade are now receiving light in intensities to which they are not accustomed. Light may provide the energy for photosynthesis, but it is also dangerous. Therefore, plants not only have methods to use light, they also have methods to protect themselves from it when it is excessive. An important part of a chloroplast's activity is related to protection from light. The two processes, photosynthesis and photo-protection, operate like cooperating valves regulated by the environment: when conditions are favourable, photosynthesis prevails, whereas when conditions become adverse, photo-protection prevails.

Phenotypic plasticity as expressed in plant biochemistry: photosynthesis and photo-protection

When there is more light than photosynthesis can handle, the excess, which now is dangerous, is directed to the photo-protective "apparatus," which renders the light harmless by turning it into heat. Therefore, a plant that suddenly receives more light than usual will try to allocate the work between the two processes. It will mobilise its biochemical plasticity by channeling resources to enhance photo-protective mechanisms. Success depends on the plant's readiness. It is a foot race between relentless light excess and the capacity of the genotype to change the conditions of chloroplast function (the biochemical phenotype) quickly, before the light destroys the photosynthetic apparatus itself. Imagine something similar in daily life: for example, a field is watered through a series of ditches supplied by a river. The extent of soil wetting depends on the water supply. If, however, the river suddenly carries an increased amount of water, the field will flood at the risk of destroying the crop. The excess water must be channeled appropriately in other directions. If the existing anti-flood mechanisms are not sufficient, there is an alert to urgently create more. What would happen to leaves if their entire biochemical and metabolic plasticity is exhausted without the results expected, that is, if the light river "overflows"?

Intensity of the stress factor and plant readiness for a fast biochemical response: a grim race

Earlier it was shown that adaptations to prevailing light intensity are not only biochemical/biophysical in nature, but also morphologic. The leaves in the shade are larger and thinner. A mature leaf, however, cannot change its size or increase its thickness, but it can always be replaced.

What if the stress factor gets the upper hand?

What human would not like to have this wonderful advantage, being able at will to replace a damaged, ageing, or underfunctioning organ with a fresh one? Among plants, if the first biochemical line of defence fails, the plant will use more drastic measures: it will activate the mechanism of rejecting the now useless leaves. Immediately afterward, using its multipotent

The advantage of organ replacement when organs are irreparably worn: if only we could do the same!

meristematic cells (see the beginning of the book), the plant will grow new leaves that will acquire properties suited to the new environment. Common experience confirms that quite often the whole aboveground part of the plant is destroyed and revived from the part below ground. Sometimes, a mere stump gives forth a robust tree.

This is how the simple nutritional demands of plants, their strong phenotypic plasticity, their capacity to create new organs, and the relative independence of their survival from the density of their population makes them tolerant to environmental changes, even if these might be intense and extreme. The combination of such properties may be quite convincing in explaining the absence of mass extinction events in the plant kingdom as well as the great number of living fossils – species that not only survived violent ecologic traumas, but also retained their external form unaltered for tens or hundreds of millions of years.

Reproductive Idiosyncrasies of Plants: Resistance to Extinction

Despite these convincing arguments, however, only half the truth has been told. The survival of individuals is an important parameter, but its ultimate efficacy for species survival depends on whether the surviving individuals are capable of reproduction, that is, passing their positive attributes to the next generation. However robust an individual might be, its significance for the species is negligible if it has an ineffectual and inaccurate reproductive system. Indeed, such individuals are dangerous to the perpetuation of the species.

Individual survival is pointless for a species if it is not accompanied by the capacity of reproduction

Do plants have reproductive attributes that give them advantages – features that can explain the absence of mass extinction incidents? Remember that after violent

Reproductive idiosyncrasies of plants and resistance to extinction

catastrophic events of global proportions, plant ecosystems might lose ground, but they always recover with more or less the same composition as before. In contrast, the dinosaurs (and many other animal groups) never recovered. How is this related to the reproductive idiosyncrasies of plants?

Hybrids are very rare among animals. Hybridisation is the cross-breeding of two organisms with different genetic compositions, from different yet related species. The organism that emerges differs from its parents. The few cases of hybridism among animals produce sterile offspring. The most common example is the cross-breeding of a donkey and a horse, which produces a mule. The mule is stronger and can endure more, but it cannot reproduce. The difficulty emerges during the processes leading to the formation of the gametes, when there is difficulty in equally distributing the chromosomes during mitotic division, because the chromosomes are not homologous. To simplify, every somatic cell of an organism contains a stable number of chromosome pairs, which contain the DNA. In humans, for example, there are 46 chromosomes, in pairs of similar ones (at least in females), that is, 23 pairs. In other words, cells are diploid (because each chromosome has its exact copy), and the chromosomes of each pair are called *homologous* and contain the same genes. Somatic cells assigned to produce the corresponding gametes (reproductive cells) – the spermatozoa in the testes and the ova in the ovaries – are divided in such a way as to produce two cells, each containing one copy of each chromosome. In other words, whereas somatic cells contain the genetic information in two copies (diploid), spermatozoa and ova have only one copy (haploid), that is, 23 chromosomes each. The diploid condition is restored when the spermatozoon and the ovum fuse to create a new organism.

Hybrids might have an interesting phenotype, but they cannot reproduce it

Homologous chromosomes and hybrid sterility

The sterility of hybrids is a result of the fact that the chromosomes of their somatic cells are not homologous; therefore, it is impossible for the genetic information to be distributed equally to the gametes during meiotic division. Therefore, although hybrids may have interesting attributes and are stronger organisms because they combine different capabilities, as far as their contribution to the next generations, they are an evolutionary cul-de-sac.

The reproductive difficulties just described do not concern only animals. Yet among plants, these difficulties may be circumvented over time and a sterile plant hybrid might become sexually fertile again through two typical plant properties: the capacity for asexual, vegetative reproduction by creating clones and the phenomenon of polyploidy. Vegetative reproduction is the creation of offspring from somatic cells, without the intervention of gametes. There is no fertilisation or gene exchange, but part of the plant develops into a new plant, which separates from the mother plant or remains attached to it. It is a bit like the creation of Eve from Adam's rib. In the case of strawberries, from time to time they grow runners that give new roots some distance away that grow into proper plants. Attachment to the mother plant is sometimes severed and sometimes not. The new plant is genetically identical to its mother; in other words, the new plants are clones. Cloning in animals is not only unknown in nature, but extremely difficult in the laboratory as well. As the famous Dolly experiment proved, the sheep created after many attempts at cloning was feeble and did not survive as long as sheep born the normal way. Animal cloning – and particularly human cloning – also raises several bioethical and legal issues, as does every human scientific effort that attempts to violate natural laws. For reasons explained elsewhere, in the discussion on the mechanisms of evolution, nature does not favour uniformity. Why is it, though, that what nature denied to animals it generously conceded to plants? More than 40 % of plant species

Plants have their way, though: vegetative reproduction and cloning

Nature denied cloning to animals but offered it generously to plants

are capable of natural cloning, that is, producing genetically identical offspring from somatic cells. The reader must already realise that this capacity is related to the multipotency of the meristem, the plant tissue made up of cells that remain undifferentiated throughout a plant's life and produce new skin, vessels, leaves, flowers, and fruit. Often, meristematic cells also produce whole new clone-plants.

More than 40 % of plant species create clones

Why is natural cloning so common among plants? It seems that what lifts the obstacles for cloning is the plants' great phenotypic plasticity. The environment of plants is particularly diverse. For example, strawberries like light, but if the maternal plant produces a runner to create a clone, the new plant might find itself in the shade of a nearby plant of another species. Because the clone is identical to the mother plant, which lies in the light and has made the necessary adaptations, the young one would have less chance in the new, shady environment. However, its phenotypic plasticity allows its identical genotype to create a "shade" phenotype. Moreover, if the stem that connects the mother plant to its descendant is not severed, the two plants will cooperate to better exploit both environments. The mother plant might photosynthesise more easily, whereas the descendant would enjoy moister soil. Therefore, the mother supplies the descendant with excess photosynthetic products, while the descendant provides the mother plant with water through the runner, which functions like an umbilical cord. Without phenotypic plasticity, the strawberry's spread over new ground would have no chance of success and would not be reinforced by natural selection as a mode of reproduction.

Lack of motility, phenotypic plasticity, and environmental diversity, or why cloning was favoured among plants

When they have phenotypic plasticity, clones exploit their environment better

It may seem contrary to common experience for so many plant species (40 %) to use this mode of reproduction as well – the term *as well* is used here because vegetative/asexual reproduction complements sexual reproduction, which is based on the normal exchange of genes via the gametes. Therefore, all plants reproduce sexually and some occasionally use vegetative reproduction as well. Why is this true for as many as 40 %, however, and why does it go unnoticed?

Vegetative reproduction and sexual reproduction are parallel processes among plants

Often, reproductive shoots lie underground in obscurity. Nobody would doubt that a cluster of poplar trees comprises as many individuals as the trunks seen; however, this may not be true. Auxiliary stems may begin at the roots of certain trees and create a second tree a few meters away. Genetic analysis often proves that even a small forest made up of such trees might, in effect, be only one individual (if the trees communicate underground) or many genetically identical individuals originating from the cloning of the same parent (if the bridges of subterranean communication have been severed). Other methods of vegetative reproduction that may go unnoticed include the production of bulbs (when one bulb produces several, and they become independent from the parent plant and give new plants) or the production of small shoots with a loose connection to the rest of the plant (as in some cacti, the thorns of which are caught in the fur of passing animals that drop the shoots in a different territory). In other cases, the new organism is produced without pollination from ovaries that have not undergone meiosis of their chromosomes (and therefore are diploid), whereupon the seed produced is genetically identical to the mother plant. This phenomenon is called *apomixis*. Finally, some cases are obvious, as in the plants of certain species of the Crassulaceae family; in these cases, tiny plants – full with leaves, a stem, and rootlets – are produced on the leaves of the mother plant, eventually dropping and creating new clone-plants.

Vegetative reproduction goes unnoticed: when the forest hides the trees

What are the advantages of vegetative (asexual) reproduction? Again, these advantages are totally different from what a human might think based on his or her experience. Childless people desperate to have a baby would not refuse the opportunity to have a child

Human cloning raises bioethical dilemmas. Among plants, it resolves practical issues.

by cultivating some cells from their tongue or skin in a test tube. If they were vain, this option might even be more desirable: the child would be as beautiful and intelligent as the parent. In nature and the world of plants, vegetative reproduction resolves much more interesting practical issues. It frees parents from the obligation of finding a mating partner. It is useful for plants that grow in isolation with no plants of the same species nearby, as is the case after an ecologic catastrophe. The same natural disaster also might destroy natural pollinators, usually insects and birds, organisms much more vulnerable than plants to ecologic trauma. Under such circumstances, vegetative reproduction is the only option. Furthermore, vegetative reproduction facilitates rapid expansion over a territory. A clone grows much faster because it continuously is provided with raw materials from the mother plant, whereas a seed must find support in the minimal reserves of the cotyledons. If, therefore, an animal and a plant travelled to an uninhabited world without any organisms, the animal would not stand a chance. The plant, having the privilege of asexual reproduction, soon would gain ground. If Noah had known about this capacity of plants, it would have saved him a great deal of space in the ark.

Reproductive self-sufficiency, or how to perpetuate the species free of the obligation of finding a mating partner

How Noah might have saved space in the ark

Asexual reproduction, of course, has the disadvantage of not achieving the flow of genes from one individual to another, so it does not pass on extra attributes or increase diversity within the species, because clones are genetically identical. For that reason, asexual reproduction coexists with sexual reproduction, and it becomes particularly important at times or in regions of great environmental pressure, when the population declines. For example, a birch tree reproduces sexually in northern Europe, which is the main territory in which it thrives. However, in the northernmost limit of its expansion, just beyond the Arctic Circle, it uses mainly asexual reproduction.

The environment determines the relative contribution of sexual reproduction and cloning

In conclusion, the capacity for asexual reproduction allows isolated individuals to survive and, combined with the typical longevity of numerous plants, offers the possibility of overcoming long-term and extreme ecologic disasters. If one also considers the capacity for hybridism between related plant species, one can easily realise that even under such difficult conditions some diversity might be achieved. Although hybrids are sexually sterile, they can reproduce asexually and extend their lives until more favourable conditions prevail. What happens if this takes a long time? How can hybrids acquire the capacity for sexual reproduction? If this becomes possible, then through gene mixing of different individuals, new attributes will emerge that will be added to those of the hybrids. Remember that a hybrid, as a union of genetic material from two different species, differs from its progenitors.

As was already explained, the sterility of hybrids is a result of their lack of homologous chromosomes. Having a line of chromosomes from each parent (each parent belongs to a different species), they cannot distribute their genetic material equally when the cells destined to produce gametes (i.e., pollen grains and ova) are divided. Among plants, however, the phenomenon of polyploidy is quite common. What is polyploidy, and why is it important?

Chromosome and gene copies: diploid and polyploid organisms

It already was stated that chromosomes in plant and animal cells are found in pairs – in other words, organisms are diploid. During cellular division, the chromosomes split from their partner and each one is duplicated, resulting in two pairs again. At this phase, every cell has four copies of each chromosome and, therefore, four copies of each gene. In humans, for example, a cell in a state of rest has 23 pairs of chromosomes (i.e., 46 chromosomes in pairs of similar ones) and it is diploid. At the specific division phase, it has 23 foursomes (i.e., 92 chromosomes in similar foursomes). This is a temporary stage. Twenty-three pairs go to one pole of the nucleus, and the other 23 to the other pole. The nucleus is divided, then the cell is divided so that two similar cells are created, with the genetic material exactly allocated to each one. Sometimes chromosomal division occurs without nuclear division, whereupon a tetraploid cell is created. This situation is not tolerated among animals, and such a cell usually dies. If such "errors" are made in animal gametes, the organism produced by sexual reproduction dies immediately. Even if the error concerns only one chromosome, the organism produced will be problematic. Down syndrome among humans is caused by the mere presence of three copies of chromosome 21, and individuals with this characteristic genetic condition usually do not reach reproductive age.

Among animals, polyploidy is forbidden

In plants, however, polyploidy is well tolerated and 50 % of plant species are polyploid. Indeed, because there are various ways for the number of homologous chromosomes (copies) to be increased, the number of copies is not always even. Furthermore, because there might be repeated episodes of increases in the number of copies, there are plant species with many tens of copies of the same chromosome per nucleus. Therefore, they might have tens of copies of the same gene per cell.

However, 50 % of plants present polyploidy

The importance of polyploidy in the evolution and long-term (in geologic terms) survival of plant populations is huge. Polyploid species are particularly resistant to lethal mutations. When one carries many genes for the same purpose, the harmful mutation of one is "diluted" within the crowd of the other, healthy ones and the likelihood of its passing to the next generation is almost eradicated. Polyploid species, therefore, are resistant and stable. On the other hand, small, harmless mutations in some of these multiple genes create a mosaic of gene products, attributing great adaptive flexibility to polyploid plants. If, for example, the gene product is an enzyme that completes a certain chemical reaction, then it is possible the enzyme in

The advantage of multiple copies of a gene, or how to blunt the impact of unfavourable mutations

question might exist in multiple forms and each form would have slightly different attributes. Therefore, one part of the enzyme population might function better at low and another at high temperatures (and, therefore, potential habitats) the plant can tolerate.

Polyploids as gene pools for future use

Returning to the initial question: How can polyploidy turn a sterile hybrid into a fertile one? An episode of polyploidy, that is, the duplication of the chromosomes of a hybrid without nuclear division, creates a nucleus with pairs of similar chromosomes. If this occurs in the cells of the reproductive organs, then the obstacle blocking the production of fertile gametes (pollen grains and ova) is removed and sexual reproduction is possible again. Therefore, hybridism combined with polyploidy is a significant driving force for the creation of new species and their establishment within the ecosystem.

Polyploidy and speciation

A final question is pending to fully explain plant resistance to mass extinction. Assume that an ecologic catastrophe leads to the disappearance of all vegetation from a vast area. How can vegetation of a similar composition reappear in this area after several tens or hundreds of years? The key seems to lie in another special attribute of plants: the dormancy and resilience of their seeds and spores. These reproductive elements mature on the mother plant and are equipped with dormancy mechanisms. This interesting issue is presented in more detail in Chapter 5. For now, *dormancy* is defined as a state of suppressed metabolism, which helps an organism face a stressful period. Although this phenomenon is not limited to plants, in the case of plant seeds and spores, the dormancy period is very extensive. Seeds remaining in the soil 10, 20, or even 100 years before sprouting and growing into healthy plants is not a rare phenomenon. The record is held by some sacred lotus (*Nelumbo* nucifera) seeds collected from the bottom of an ancient lake, dated, and found to be around 1,000 years old; 60 % of these

Dormancy and resilience of seeds and spores

seeds sprouted normally and grew into healthy plants. Such extraordinary capacity for seeds to resist degradation is the result of several factors, such as walls impermeable to water and O_2, mechanical strength that provides the walls with the ability to resist microbial attacks, the presence of protein-repairing enzymes inside the seed, the capacity of cellular membranes to retain their fluidity, the high concentrations of reductive substances that prevent oxidation, and the supply of hormones that suppress metabolism. Similar strength capacities have been recorded in cases of bacterial and fungal spores, but not among animal organisms at any phase of their development.

Descendants wait underground for the right moment

Chapter 5
Sex in Nonmotile Organisms

Hermaphroditic in Form, Yet Sexually Segregated with the Help of Chemistry

Flowers are disposable sexual organs

Flowers are a plant's genitalia; when people cut flowers, they actually are cutting the plant's genital organs – an atrocious act from the plant's perspective. However, a plant rarely has only one genital organ, it usually has many; but even if a plant had only one genital organ, it would grow another one the following year. Therefore, these genitalia are disposable, although this is not an excuse for overpicking flowers. If a plant species is threatened by extinction, removing even one of its flowers might be critical.

Plants are usually hermaphroditic. However, they avoid self-fertilisation as much as possible

These organs guarantee plants' reproduction. Most plants are hermaphroditic, that is, the same flower serves both male and female functions. Plants, therefore, are capable of self-fertilisation, but only in rare cases. Because the goal of sexual reproduction is to mix genes so that the offspring produced is slightly different from its progenitors, self-fertilisation is not favoured. A population composed of identical individuals, such as those produced by self-fertilisation or cloning, would be less robust. In case of an environmental change, all individuals within this population would have exactly the same potential to face the new situation. If this environmental change were unfavourable, none of the individuals would have an advantage over the others. On the other hand, sexual reproduction results in a medley of qualities among individuals in a population, ensuring that some of the individuals will be more likely to resist and survive unfavourable changes and to leave behind more descendants, thus increasing the frequency of the genes making their owners more resistant within the population.

What are the advantages of sexual reproduction?

How is self-fertilisation avoided? In certain plants, the same individual produces two types of flowers, male and female, in different positions. Another group of

Are there male and female plants?

plants (about 10 % of all plant species) have separate male and female individuals, that is, those that have exclusively male or exclusively female genital organs (flowers). Indeed, in most cases, the morphologic difference between male and female organs is quite obvious, as it is in humans. Sometimes there are secondary features that distinguish each sex, but these are not as pronounced as in the animal kingdom. For example, the male lentisk (*Pistacia lentiscus*, or mastic tree) is taller than the female. Lentisk pollen is airborne, which means the wind helps the male sperm cells in the pollen grains to reach the genitalia of the female plants. Therefore, for fertilisation to take place, the female's flowers are located beneath those of the male so as to receive the pollen grains before they fall to the ground.

Sexual organs at a distance

The case just described is a simple one, but how do hermaphroditic plants avoid self-fertilisation? If one had both a penis and a vulva and no moral inhibitions, one could easily and self-sufficiently make love to oneself. Of course, this would presuppose that the two sets of organs faced each other at the proper projectile distance. Imagine, though, that the penis is situated at the neck and the vulva at its usual location. Despite one's desire, imagination, and inventiveness, there would be no position to enable intercourse. Therefore, one way to avoid undesirable self-fertilisation is spatial separation of male and female functions on plants. To make this concept more understandable, a brief description of a flower and its functions is required.

What does a flower consist of?

A typical flower has (usually) green sepals that look like small leaves; these protect the base of the flower and, through photosynthesis, contribute to the growth and maintenance of the flower as well as to nectar production. The petals also resemble leaves but are not green, except for the odd exception. The bright petal colours attract pollinators, as is described later in this chapter. The male function is performed by the stamens, which bear the anthers at the top of their

The basics of reproductive function

filaments. The anthers produce pollen grains, which contain the male reproductive cells. The female function is served by the ovary, usually distended at its base, which contains the ovules; the stylus begins at the ovary and ends with a stigma. Pollen grains – carried by the wind in certain plants (anemophilous species) or transported by animals (usually insects) in others (entomophilous species) – land on the stigma. The surface of the stigma is sticky, which helps hold pollen grains. When a pollen grain finds itself on a receptive stigma, it sprouts and creates a tubule that enters the stigma, penetrates it, and travels down through the stylus towards the ovary, where it approaches an ovule. The male reproductive cells are released and fertilise the ovule.

Ovule fertilisation triggers the creation of a seed, which comprises the embryo and the endosperm. The seed is surrounded by a hard perisperm, which serves a purely mechanical protection role. Usually, the ovary contains many ovules, so numerous seeds are formed. When all available ovules have been fertilised, the petals age, fade, and drop, while the ovary walls and other parts of the flower swell and create what is known as a fruit. The stones or pips, which are usually at the centre of the fruit – for example, in the core of an apple– are the seeds. The same may be said about the olive, which, however, has only one seed, its pit. Many fruits are small and colourless; others are sizeable, fleshy, and colourful. All of them, however, have a mechanism to disperse their seeds. These mechanisms ensure that the new plant will not establish itself near its parents. In any case, parental care is not very common among plants and the parental contribution usually ends once the seeds (whether packed within a fruit or not) leave the maternal plant. Any parental care involved has ended in the previous stages – and quite successfully. Seeds have dormancy mechanisms to prevent them from sprouting immediately, as the dispersal season may not be the best

Why should descendant plants not sprout immediately?

for sprouting. For example, it would lead to failure if an annual plant of the Mediterranean region that disperses its seeds in the spring sprouted immediately. The young plant would not be equipped to face the coming summer drought. The dormant period ends when the seed's sensors detect suitable environmental conditions that ensure the likelihood of the new plant's survival. The seed also is equipped with a tough and resistant shell for mechanical protection and a potent biochemical antioxidation mechanism. It may take years, decades, or even centuries before a seed finds the right opportunity to sprout.

Why should descendants distance themselves from their progenitors?

Dispersal mechanisms are important for two reasons. First, they ensure that the new plant will not establish itself near its parent and will not have to face parental competition for available resources. It is not in the young plant's interest to grow in the shade of its progenitor. Second, dispersal means expansion of the species' territory. The various mechanisms of dispersal and their importance to the geographic spread of plants are examined later.

Now that a morphologic description of the flower has been given, the ways hermaphroditic plants avoid self-fertilisation can be analysed further. One method is for male and female flowers to grow in different locations on the same plant; about 10 % of plants have this type of dimorphism. The male flower is incomplete, that is, it bears only stamens and produces pollen, whereas the female flower (also incomplete) bears an ovary and stigma and produces ovules. The individual, therefore, is a hermaphroditic organism, but the male and female organs are at a distance from each other, as in the imaginary human example presented earlier. However, human reproduction could still be carried out in a self-sufficient manner by conveniently bringing into contact the products of the male and female reproductive organs through another, nonsexual organ. This is even easier among plants, in which penetration by the male organ into the female one is not necessary; all that is needed is for pollen grains to be carried to the stigma with the help of the wind or insects. In other words, spatial separation is a halfway measure that contributes somewhat towards preventing self-fertilisation. Even more effective is temporal separation, in which the male and female functions of the same individual do not mature at the same time, but hours or days apart. This way, the landing of a pollen grain on the ovary stigma of the same individual (the ovules of which have not matured yet) is fruitless. Because the individuals of the same species within the population are not synchronised in the maturation of their corresponding organs, there will always be a mature ovary nearby to receive a neighbour's pollen but not the pollen of its cohabitant.

Why do the male and female flower functions mature at different times in many plant species?

When females do not accept any old Tom, Dick, or Harry

What happens, though, when the male and female parts are not separated spatially or temporally? How is self-fertilisation avoided in hermaphroditic plants bearing hermaphroditic flowers with simultaneous maturation of male and female functions? In this case, the safest exclusion method – biochemistry – comes into play. The question is whether the stigma can recognise the origin of the pollen grain that has landed on it. In deciding whether to accept the grain, the stigma must discern whether the grain is from a neighbour or from itself – or even worse, whether it is from a strange creature that does not belong to its own species. No female wants to sacrifice her ovules to a fool's errand in unnatural actions. Pollination is difficult with so many libertine pollen grains wandering in the air and another lot arriving by regular delivery via insects. On the other hand, of course, the pollen grain also has begun a journey of unknown destination. It is equally interested in where it will end up and if its labours will result in a serious outcome – after all, it carries a packet of genes it is proud of and would like to hand down to the next generation at all costs. Its success, however, depends on

When sex is in the air, biochemistry intervenes

unexpected changes in the wind and on whether the delivery service (i.e., the insect) knows the right address.

A DNA test before penetration

The technical term for biochemical pollination blockage is *incompatibility*. In essence, it is a comparison of the gene products of the stigma with those of the pollen grains. If the products are identical, then the pollen comes from the same individual. If they are very different, then it comes from a different species. In either case, the comparison trial gives a negative verdict: the visitor does not carry the right passport to be allowed in, because it will cause trouble. It is better for it to stay outside – why should an importunate gatecrasher be allowed to sprout a penetration tube? In these cases, therefore, the pollen grain does not sprout, as if it has no desire to penetrate the stigma with all the disturbing chemicals it is emitting. In other cases, the pollen grain may sprout and its tube may begin penetrating the stylus. At this point, however, a second control point is activated: if the visitor comes from the same individual, then the pollen grain tube opens and releases the genetic cells prematurely, before they arrive at the proper position, that is, near the ovule. In other words, it is something like premature ejaculation.

Where there is crowding, the neighbour benefits

Even if the previous chemical control systems of the intruder's identity prove ineffective, there is a third obstacle: the penetration speed of the neighbour's tube is greater than that of the cohabitant's. In other words, at the scrimmage and the hubbub, the neighbour is at an advantage and the first to reach the final target. Therefore, in the case of appropriate similarity and compatible chemistry, the two partners become aroused and mutually receptive; then, the pollen grain sprouts and the stigma relaxes and accepts the penetration by the pollen tubule. The latter goes in deeply and releases its genetic cells, which will fuse with the ovules to create seeds.

A strange yet effective system of immunity control

The incompatibility between pollen grains and the stigma brings to mind the function of the human immune system. In the latter case, any foreign cell entering the organism is checked, and if it does not have the necessary immune identity, it is rejected. Indeed, in the case of pollen from another species, the phenomenon is similar. However, in the case of pollen from individuals of the same species, things are reversed: the cells of the same individual are rejected whereas entry is allowed to cells of other individuals of the same species. In this sense, the system of cellular recognition in the plant kingdom is unusual.

Solitude and Self-pollination

There are times when plants self-fertilise

Cellular recognition and self-pollination prevention systems in plants, however, are not fully effective. Throughout this book, it is stressed that growth, development, and behaviour among plants have a characteristic plasticity. There are species that demonstrate almost absolute avoidance of self-pollination. At the other end of the spectrum are plants in which self-pollination is quite frequent. Between the two extremes, there are several intermediate cases. Absolute incompatibility is sometimes extended to include individuals that are "relatives." For example, if one examines the possibility of being fertilised by all the other individuals of a population, one might discover that not only self-fertilisation but also cross-fertilisation by a percentage of one's neighbours is impossible. In a genetic analysis of the members of a population, one would discover that the individuals that do not reproduce using each other's genetic material are related. In intermediate cases, one might find plant species with a higher or lower number of seeds originating from self-fertilisation. Some hermaphroditic plants bear two kinds of flowers on the same individual. Some are large, brightly coloured,

Organs to serve the neighbour and organs for self-fertilisation

and located conspicuously to attract the interest of insects; these have a high incompatibility between pollen grains and the stigma, that is, they are flowers that exclude self-fertilisation. The other flowers are small, located inconspicuously within the foliage or even underground, and do not have lovely colours or produce nectar. These flowers do not even bother to open in some cases, and they are self-fertilised at the bud stage.

What, then, explains the cases in which barriers (spatial, temporal, and biochemical/genetic) collapse and species self-fertilise conveniently without any restrictions, and when does this happen? Does self-fertilisation, under certain conditions, have advantages that outweigh those of cross-fertilisation? Keep in mind that mixing genes in sexual reproduction aims mainly to create offspring with a diversity of characters, to ensure the population becomes more robust.

Why does solitude favour self-fertilisation?

Imagine someone shipwrecked on a desert island. However much that person might want to disperse his or her genes and become a parent, it would be impossible and the future human presence there would be doomed. What is possible for a plant is not the case for humans. A plant can perpetuate its species in the following ways: If it is a plant with self-fertilisation potential and loose compatibility control, the problem is solved. If it has full control of compatibility, it can proceed patiently with vegetative reproduction (i.e., creating clones; see the related discussion on the reasons mass extinctions of plant species were prevented in the past) and survive for centuries in that state until a rescuing mutation loosens incompatibility and restrictions. In this way, the plant might join the self-fertilised group, enriching the repertory of its reproduction. Obviously, this manner of reproduction produces clones. However, random mutations will provide opportunities for variability within the population and the possibility for a gradual return to a state of cross-fertilisation. Therefore, self-fertilisation is advantageous for individuals living in isolation or within a sparse population.

Now imagine that a man is stranded on an isolated island and discovers that a beautiful woman has found refuge on a nearby island. Unfortunately, neither can swim. However much the man racks his brains, he cannot find a solution. In effect, the situation is the same as that of two exclusively cross-fertilised plants standing next to each other. Burning with the flame of unfulfilled desire, he might choose some sort of compromise, a go-between, just as plants do: "Come here, lad; this is for your trouble. Deliver this letter to the lady across the water." This is not to imply it is the same thing, but it is better than nothing. However, what if the lad becomes lost, or is killed along the way, or goes on a long-term strike?

Self-fertilisation might be just the solution when times are tough…

Therefore, it is obvious that self-fertilisation is an advantage not only in cases of solitude, but also when pollinators are scarce. As is shown later, plants' flowering periods are synchronised with the activity of pollinators. In any case, such intermediation is not without gain, because both partners benefit from the relationship. However, there are ecosystems in which the density of potential pollinators is low, such as areas with permanently strong winds or very low temperatures. Furthermore, in other places that are generally favourable for insects, their numbers might fluctuate significantly from year to year because of epidemics, predator attacks, or transient bad weather. In these cases, the alternative – self-fertilisation – ensures a minimum number of descendants, even during difficult times.

…and it also costs less

Self-fertilisation does not require the participation of insects; therefore, the plant does not go to the expense of creating large and conspicuous flowers or of equipping them with colours, scents, and nectar. Sexual reproduction comes at a cost. Therefore, self-fertilised plants produce debased flowers with regard to size and to the goods and services provided to pollinators. However, plants never completely give up on producing flowers of certain specifications, which, with small alterations, will be useful in attracting insects or other animals when better days come. Self-fertilisation, like vegetative reproduction with clones, is just one more expression of plant plasticity, one more strategy to circumvent the risks of reduced offspring and to increase the likelihood of survival for individuals and populations under stress. Furthermore, it offers these populations the possibility of spreading quickly in new territories and environments. They do not rely on insects to ensure their reproduction. When a region is cleared of its established organisms by a natural disaster or human intervention (which today is a natural disaster), the remaining plants soon thrive in the new territories, using, among other things, their colossal, flexible, and particularly plastic reproductive potential. Animals always lag behind.

As Always, Heterosexuality Carries a High Price; However, in the Case of Plants, it Is the Middleman that Profits

Plants do not choose their mates

In further studying plant reproduction and the organs that perform it, let us leave aside the solitude of self-fertilisation and focus on sexual reproduction. A sedentary plant cannot choose its mate. However, like all other organisms, it tries to disperse its genes to the next generation in the best way possible; this way has its cost. In exactly the same manner and with similar hopes, humans go to the gym, seek wealth and social status in fair or unfair ways, try to improve their physical image with expensive brands of clothing and impressive cars, cultivate virtues or "virtues," and become obsessed with "appearing" and occasionally with "being," so as to attract the attention of the opposite sex. This also carries a high price but has its rewards. The plant has no reason to seduce the opposite sex, but it must appeal to the insect, the delivery lad, and must pay the price. No insect would carry pollen because of compassion, support, or understanding for the plant's problems. The cooperation of insects (or other animals) with plants – which results in pollination and reproduction for the latter – is a (mostly) mutually beneficial transaction.

Flowers are pretty to appeal to insects not to seduce us or other plants

Some relevant questions are raised: What is the benefit for the insect? Why should it approach the flower? Before these questions are answered, it must be noted that not all insects are pollinators. On the contrary, only a small minority engages in this transaction with plants. Most insects consume plants without offering anything in return. The reward for pollinators, therefore, is food – high-quality food in the form of pollen and nectar. Nectar is manufactured by plants exclusively as a lure and a reward. It is a concentrated (15 % to 75 % by weight) water mixture of sugars similar to syrup, and its main ingredients are glucose, fructose, and sucrose. The ratio of the three sugars varies from plant to plant and usually depends on the preferences of the main pollinators of the specific plant species. Besides sugars, nectar also contains small quantities of amino acids at a level of around 1 %.

An honest deal with mutual benefits

Nectar as a reward

Although nectar is not nutritionally complete, it has a high energy value. Butterflies, which are the final developmental stage of certain insects, have minimal nutritional demands. In effect, the only thing they seek is fuel for their flight, as they die right after intercourse and egg laying. Therefore, during their short lives, they feed exclusively on nectar. Nectar is produced by the photosynthesis of leaves situated near the flowers. From these leaves, plant vessels transport the nectar to special formations called nectaries located at the base of the

One must try if one wants the reward

flower. Readers may have seen a butterfly dip its long, thin straw-like proboscis inside a flower. In its effort to suck up the nectar, it is sprinkled with pollen, which it transfers to the flowers of the next plant it visits. Of course, this presupposes that the stamens are in the right position, that is, somewhat blocking the butterfly's path towards its food. Therefore, the nectar usually lies at the base of a flower and the stamens high up at its entrance. That way, even smaller insects interested in the nectar may serve as pollinators, because as they enter and exit the flower, they are bound to touch the stamens. Generally speaking, the morphologic traits of plant flowers and of the bodies of insects serving as pollinators have been subjected to the same selection pressure, and they coevolved to match the functions to be performed. In other words, plants that at some point made certain modifications to the position of their nectaries and anthers to improve their contact with insects were more likely to have more numerous offspring and therefore to stabilise the new flower form in their population. Similarly, insects that through various mutations modified their body design to match the shape and size of a specific flower enjoyed more food and likewise produced more descendants. The final outcome of this long evolutionary process is specialised relationships in which specific insect (and other animal) species pollinate specific plant species for their mutual benefit.

The rules of the game

The form of the flower and the body features of the pollinators coevolved to match each other

Of course, nature has its con artists too. Many insects are small enough to enter deep into the flower and drink the nectar without ever touching the stamens. Others, even cheekier, prick the base of the petals and suck the nectar, never going through the proper entrance. As in human society, there are all sorts of characters in nature, from honest professionals who observe the rules of the game to dirty snatchers. One should not think, however, that such cunning is typical only of animals. Many plants play a similar crafty game. Some of their flowers do not produce nectar, so an animal might leave

There are crook insects...

...and crook plants

"empty-handed" and have to search for better luck elsewhere, but it has already filled up on pollen. Obviously, the plant keeps its nectar production costs low this way. Other impressive tricks will be demonstrated later during this exploration of the unending struggle for survival.

Bees: the perfect pollinators

The most common pollinators are bees (wild and "bred") and butterflies as well as beetle, fly, bird, and bat species. For now, let us focus on the bees. Bees collect nectar and pollen; the nectar fuels their flight, whereas the pollen is stored in sacs on their hind legs, which are visible to the naked eye. The bees transport the pollen to the hive to be used as food for the queen and larvae. Not all the nectar is used up in the bees' flight, and they bring the leftover nectar inside their hive to use as a food supplement, converting it into honey. Quantitative data on honey production are impressive: bees produce 1 kilo of honey by visiting around 17 million flowers. If this task were assigned to only one bee, it would need to fly for 3 years. As for the plant, a medium-sized bush with a projected surface area of 1 m^2 would need about 2 years' worth of photosynthesis to produce the sugars contained in a kilo of honey. One kilo of honey costs €4 to €5, a truly low wage for so many years of hard labour by both the plant and the bee, and for a product of such superb quality.

Why is honey cheap?

Although nectar is a high-energy food, it is far from complete. The amount of nitrogen it contains is too low to meet the nutritional requirements of animal organisms. Furthermore, it lacks minerals, lipids, and vitamins. Therefore, as wages for the reproductive services provided by insects to plants, it is lacking. The insect needs to supplement its diet from other sources, which means it probably will detour from the expected destination, the next flower to which it will carry pollen. There is no point in the bee visiting a protein (nitrogen) source (i.e., some excreta) in the meantime and leaving the pollen there. Therefore, although evolution selected nectar as an exclusive reward for pollinators and nectar has no other function for plants, there was pressure for the production of a more complete food that would better close the deal. That way, the intermediary would have no reason to take on a second job to supplement its livelihood. The solution selected was to share part of the product to be carried, that is, pollen. This solution has come at a high price for plants, but as explained earlier, reproduction is a top priority.

Towards an improved reward

Pollen is indeed a complete food; it is 15 % to 30 % protein, 3 % to 10 % fat, and 1 % to 7 % carbohydrate, and also contains a wealth of vitamins and mineral nutrients. It might not be considered a delicacy, but it is of particularly high nutritious value, although this value has no relation to its reproductive role. Plants of species with airborne pollen do not enrich their pollen with all these goodies that have no direct benefit for the successful reproduction of the plant. The extras are merely a reward for the go-between. Additional nutrients are found at the surface of pollen grains, and their usually yellow-red colour is a result of the presence of carotenoids and flavonoids. Carotenoids are precursors of vitamin A; they are potent antioxidants, as are flavonoids. Therefore, nectar is believed to have been invented originally as an exclusive nutritional enticement and a reward for pollinators, which, however, were not satisfied with this specific pay. Therefore, the deal was improved by the qualitative and quantitative enrichment of pollen, part of which is sacrificed for the benefit of the intermediary. The latter now has less incentive to visit other food sources, contributing to a more loyal relationship between the two parties.

Pollen as a supplementary reward

Of course, the cost for plants is very high. Not only must they produce more pollen, because a significant part is given not to the actual beneficiary but to the matchmaker, but they also must add expensive frills. However, it is worth the price. Plants with airborne pollen, which do not rely on living intermediaries that act – as far as finding their target goes – like "smart bombs" (later it will be shown how plants use pollinators as smart bombs), must rely on the caprice of the wind. There is a difference between using smart bombs and using hit-or-miss techniques. Indeed, in the case of airborne pollen, pollination is a random affair and the only adaptation that seems to improve things somewhat is that the release of pollen grains usually occurs when the wind is blowing and rarely when there is no wind. Those who suffer from allergies know that things get worse on windy days.

Pollinators act like smart bombs

Imagine you are a plant whose pollen is carried by insects; you are of medium size and make a deal with a carrier firm. When the carrier asks you who the recipient is, you say, "Anyone with flowers of the same shape, colour, size, and scent." You also state that part of the fare will be paid by the sender and part by the recipient upon delivery. It would be stupid, therefore, for the carrier not to conclude the job so as to get fully paid. By doing the math – that is, how many pollen grains are to be transported, how many will be dropped on the way, how many of those manufactured will be blanks, how many will be pointlessly delivered on your own flowers, and how many need to get to the female plant over there so that your dignity is not totally in shreds – you conclude that you have to manufacture, say, 500 pollen grains per flower (the number is realistic). However, if you dislike intermediaries and want to rely on your own resources – that is, the wind, which asks for no reward – you would need to manufacture at least 50,000 pollen grains per flower if you do not want to remain childless. If you calculate that during the flowering season (i.e., when the plants are "in heat") you will produce, say, 100 flowers, this would easily add up to several million pollen grains, despite your medium size. Of course, the cost will not be proportionate to the increase in the number of pollen grains.

When reproduction relies on going with the wind

When its pollen is airborne, a plant does not need to provide it with additional nutrients, especially since the excess weight would soon bring the pollen down to the ground. Such plants do not produce nectar and do not need to advertise the presence of their flowers with colours and scents. The cost of advertising and transportation is necessarily shifted to the production of a large number of pollen grains. Indeed, such plants have small colourless and scentless, almost obscure, flowers. However, their pollen production is impressive. It is visible to the naked eye in the spring, when yellow films cover the windscreens of cars or float on water, as well as in the pollen shower flying from pine cones when the wind blows. Special pollen traps have helped us calculate that within ecosystems where plants with airborne pollen prevail, 300 million pollen grains land every year on each square meter of ground area. All plant families include such species, and conifers and grasses almost exclusively comprise plants whose pollen is airborne.

How far can the wind carry pollen?

Wind transport offers a wider range with regard to the maximum distance over which pollen can be carried. For example, pollen of the common northern pine has been found on Spitsbergen Island, which is situated 750 km north of the northernmost point that species reaches, whereas for other conifer species, distances of several thousand kilometres have been recorded. Transport over long distances, however, is not effective, because the further away from the pollen dispersal point (e.g., from a pine cone), the lower the pollen density and, therefore, the lower its likelihood of landing on the right stigma. Remember that the flower stigma, which contains the entrance to the ovary, is the size of a small pinhead; therefore, although the wind may transport the pollen further than an animal would, this is not necessarily effective as far as reproduction. Insects may not reach as far, but the pollen they carry is well-packed and satisfactorily dense. Therefore, the reproductive range of plant species with airborne or animal-borne pollen does not differ significantly and ranges from a few metres to a few hundred metres.

When there is a crowd of suitors

Obviously, a significant number of pollen grains land on every flower, most of which cannot be accepted either because they originate from the same individual or because they come from a different species. Within this crowd of prospective but hardly hopeless suitors, eligible candidates might lose their way because of mechanical obstruction. Various ways exist to avoid such an unfortunate event; one way is synchronised flowering. In natural ecosystems, the competition for pollinators is strong. Therefore, selection has resulted in various plant species flowering at different times or, at least, with blooming seasons that do not completely coincide. For example, rosemary flowers in winter. There may be fewer pollinators then, but there are very few other plants in bloom; therefore, competition for those few pollinators is negligible. Furthermore, because competition is low, products need not be of such high quality. When all the other shops are closed, customers will buy from the convenience store, even if the assistant is rude and the prices high.

Some insects are truly loyal: they always visit plants of the same species

A second way to avoid crowding of irrelevant pollen grains is the phenomenon whereby bees and other insects "go shopping," which is termed *flower loyalty*. When an insect leaves its nest on the way to its first visit, it surveys the territory and notes the frequency of various flowers from different plant species. If the insect observes that poppy flowers, for example, are more numerous than the flowers of other plant species, it records information in its short-term memory on the shape, colour, and scent of the poppy flowers, which hold the most promise for a warranted supply of food, at least for the next hours, days, or weeks. The insect fixes this information in its memory following a trial visit to check the accessibility of the food source (i.e., whether the flower and the insect body are a good match) as well as its quality. The insects that can communicate with one another pass on the intelligence to the rest of the individuals of the colony or nest. This way, until the poppy flowers disappear, the insects visit only them, unless they are taken in by a similar flower. Therefore, the pollen transported belongs to only one plant species and crowding among pursuers of the same stigma is minimal. When poppy flowers become scarcer and field surveillance clearly shows that poppies no longer prevail, their image is deleted from the insect's short-term memory and replaced with the new prevalent flower, provided this flower suits the nutritional requirements and body dimensions of the insect in question.

Colours and Scents

Why do plants produce colours and scents?

As explained earlier, the colour and scent of a flower both exist to help potential pollinators locate it. The first signal an insect receives when looking for food is an olfactory one, which comes from volatile substances emanating from the flower. When an insect moves towards the scent source and approaches the plant, it receives a visual signal allowing it to spot the flower against the green background of the leaves. Obviously, the flower needs to stand out, to contrast well against the green foliage, so it may be spotted. This is why flowers are usually in conspicuous locations, at the periphery or top of plants, at the end of (usually) long stems. Intense and longstanding (for millions of years) competition among plants to attract pollinators has led to the wide range of flower shapes, colours, patterns, and scents that exist today, the ultimate goal being to increase reproductive success. From an evolutionary perspective, one can see that during the geologic periods in which the biodiversity among plant species and flower types increased, there was a corresponding increase in insect biodiversity.

Chemical and visual communication between flowers and insects: colours and scents advertise the reward offered

Flowering plants and insects coevolved; in other words, changes in the form and physiology of one group affected the other; therefore, they applied a natural evolutionary pressure on each other. For example, a flower in which the distance between the petal edge and the nectary is long favours a nectar-sucking insect with an elongated mouth apparatus (proboscis). Another characteristic feature is the parallel evolution of physiologic functions. To have some adaptive value, the colours and scents of flowers should be perceived by the pollinators; this means the pollinators can perceive the particular visual and olfactory stimuli via their sensory organs. Thus, for instance, insects need light receptors in their eyes that are stimulated by the corresponding flower colour. The terms *light receptors* and *colours* imply the chemical structure that provides the particular visual qualities and the ability of the organisms involved to synthesise the right chemical structures and place them at the appropriate positions. In other words, the chemical factory that was already established as early as the initial appearance of the first bacterial cells on Earth, and that was inherited by all other organisms, improved in all the appropriate ways to produce the new necessary substances. Hence, the change in question became established as a stable and inheritable feature for the two partners. It would be meaningless to wonder which of the two partners led this race of evolutionary changes; the other one would have to adapt. If the plant follows and cannot adapt, its only alternative is to temporarily follow the lonely road of self-fertilisation and cloning, hoping for better days.

For colour to function as a visual signal, the receptor must have a corresponding chromatic vision

What chemicals colour plants?

It is noteworthy that the large palette of flower colours and shades is the result of only a few – definitely less than 10 – chemical substances. These substances are the anthocyanins – pelargonidin, delphinidin, and cyanidin – and the carotenoids – zeaxanthine, β-carotene, and flavoxanthin – appearing either individually or in various combinations. They are enough to produce the whole range of colours observed among flowers. Furthermore, slight chemical modifications to the basic chemical structures of these substances significantly increase the range of potential colour shades.

Why are some flowers white?

What about colourless flowers? To whom do white flowers cater? A distinction must be made here. The human eye is sensitive to radiation wavelengths between 400 and 700 nm. The names of various colours have been given to the perceptions humans receive from various wavelengths within this range. However, not all organisms have chromatic vision. Beetles, for example, cannot see colours; they simply see black, white, and shades of grey. Therefore, when they need to locate a flower, they rely more on their sense of smell. The same goes for pollinators that are active at night. They do not need to perceive colour, and plants relying exclusively on such night visitors for their reproduction have no need to manufacture coloured flowers. They can shift their production costs from manufacturing colours to manufacturing scents. Other organisms, however, not only have chromatic vision, but the range of wavelengths their eyes can perceive differs from that of humans. For instance, although bees do not see red, they see within the ultraviolet wavelength band, which is totally invisible to the human eye. Therefore, a flower that seems colourless to us (such as a white one) is not necessarily colourless to bees if the flower has substances that absorb in the ultraviolet region of the spectrum. Indeed, this seems to be the case for white flowers containing the right substances, the so-called flavonoids.

What colours do insects see?

What humans "see" is not necessarily what other animals "see"

Flower traffic signals

Why are some flowers multicoloured? Why do they have colour patterns, that is, sections of different colours and shades? A common motif, for example, is lines beginning at the tip of the petals and ending at the depth of the flower, where the nectar lies. Are these lines auxiliary signals directing insects to the nectary, where they will receive their reward for the reproductive services provided? Are they like traffic signals that help prevent unnecessary jams and collisions? Such lineation occurs more frequently than one may think based on observation with the naked eye, because such traffic signals usually lie in the ultraviolet region of the spectrum, which is invisible to humans.

The world of scents

Although the chemical base of flower colours is relatively simple, the same is not true of scents. The significance of scent is catalytic for the way most animals perceive the world. Humans' sense of smell, however, is not as acute, and they perceive the world mainly through the frequency and intensity of radiation (vision) and sound waves (hearing). The insect world may not be a photon and sound map – it is not a world of physics; rather, it is a map of scents, a molecular – or chemical – world. With extraordinary sensitivity, insects may perceive the chemical nature of volatile substances and draw conclusions about their concentration. Then, depending on the gradient of this concentration, they approach the scent's source or flee from it. Insects detect the chemistry of the atmosphere not only to assess a food source but also to find their mates. They communicate and locate individuals of the opposite sex by releasing volatile substances called *pheromones* into the air. Therefore, like Jean-Baptist Grenouille, the hero of the novel *The Perfume* by the German author Patrick Ziskind, insects communicate with each other through smells.

Like Jean-Baptist Grenouille, insects communicate with smells

A fundamental distinction must be made here. As is true for light, the scents to which human smell is sensitive are not necessarily the same ones that arouse the senses of other animals. On the contrary, studies of insect pheromones have shown there is a wide range of volatile substances that humans cannot smell but that play an important role in insect behaviour. As is shown later, the incredible olfactory capacities of animals have given plants the opportunity to experiment not only with original volatile substances for helping insects locate their nutritional reward, but also with substances

What we smell in the atmosphere is a negligible percentage of what actually is there

insects have used already, as a rule, in their reproductive relationships. The capacity of such experimentation among plants has been aided by the wealth of metabolic pathways – particularly those of terpenoids and phenols – that plants have developed to defend themselves against their predators and to chemically communicate with one another and with other organisms. More about this issue is presented in Chapter 7. In any case, this is how complex relationships have been established between plants and insects; these relationships not only are mutual, they also involve deception.

This book makes it clear that flowers do not produce colours and scents to please humans, nor is such production a whim of nature. Colours and scents are what advertise a plant's products, pollen and nectar. Such advertising, though, comes at a high cost. The reader may have noticed that flowers do not smell all the time, only during certain times of the day or night. The production and release of odorous substances coincides with the maturation of pollen and often follows a diurnal pattern, depending on the period during which the pollinator is active. There is no point in advertising one's products if they are not ready for sale, or in broadcasting the commercial when the target audience is asleep. For example, there are no advertisements for children's toys on television at three in the morning. Commercial broadcasts and timely production are ruled by the economy and always take into account supply and demand. This also is true of plants' scents and colours. For example, plants relying on animals that have no colour vision (e.g., beetles) or on nocturnal animals (e.g., bats) do not go to the expense of producing colours, but they do spend a great deal on volatile substances. Economising takes other forms as well: When a flower has produced its required load of pollen and all ovaries have been fertilised, there is no reason for further visits. Therefore, the production of volatile substances and nectar terminates, and all flower parts

How can the cost of reproduction be curtailed?

When pollination takes place at night

After its ovaries have been fertilized, how does a flower indicate it is no longer receptive?

aimed at attracting insects begin to age. For instance, the colours of the petals fade, which is the first indication that the flower is no longer receptive. Further attempts and visits by the pollinator not only would be pointless, but also would be bothersome to the plant. When petals lose their colour and start falling, it means "do not disturb" or "go play elsewhere." The process of petal ageing is active and programmed, and the signal for its initiation is given through plant hormones produced by the fertilised ovary. This is one of many cases in which parts of plant organs, entire plant organs, or sets of organs age, die, and sever their relationship with the plant. Remember the seasonal falling of leaves and fruit, as well as the death of the whole aboveground part, of plants that live through unfavourable periods with only their underground parts intact.

The subjectivity of olfactory perception

From the subjective viewpoint of human olfactory preferences, smells are classified as pleasant or unpleasant. This also is true of the olfactory preferences of other animal organisms. An odour that is revolting to a human may be exceptionally enticing to another organism. Anyone who ever has to relieve him- or herself outdoors makes sure to leave the scene as soon as possible, not only because of the foul smell, but also to avoid the flies, which seem to find irresistible that which repels humans. What is it that attracts flies (and other insects) to excreta? It is a series of volatile nitrogen-containing substances – products of protein breakdown – such as methylamine, propylamine, and scatole. These substances drive flies wild, because they indicate locations with decomposing proteins, which are a delicacy for these particular organisms. This fact did not go amiss among some plants, the flowers of which smell unpleasant to humans but delightful to certain insects.

Why do revolting odours drive flies crazy?

The flowers in some plants of the Araceae family comprise a green funnel-like container (called a *spathe*), at the base of which lies a long cylinder that resembles a club (it is called a *spadix*). The whole system is a composite flower or inflorescence, that is, a formation of many small flowers. The spadix is yellow and bears many tiny male flowers, whereas the females are at the bottom of the spathe. When the flowers mature, the spadix gives off a revolting smell reminiscent of faeces and rotten meat, caused by volatile amines and scatoles. Of course, neither meat nor protein is present; this is a classic case of deception. When drawn in, coprovorous and scavenging insects crowd enthusiastically into the flowers of these plants, searching for the source of this desirable odour. Not only do the victims of this deception fail to find the food they expect, they also become trapped in a sticky substance secreted by the interior walls of the spathe. The following night, the prisoners are sprinkled with the pollen that drops from the mature stamens. In the morning, some spathe cells lose water selectively and shrink in relation to those above and below them, whereupon the relief of the spathe interior wall becomes more pronounced; this way, a step-like formation is created allowing the prisoners to escape using this "flight of stairs." The insects' adventure is over, and they can now use their sense of smell to search for real food or be taken in again, thus pollinating the flower of another plant of the same species. This strange story has another aspect: the volatility of amines and scatoles increases with the temperature. It is well known that the smell of faeces is more intense just after leaving the body, when they are still fresh and warm. If one relieves him- or herself outdoors on a frosty morning, the smell would not be as intense as if the same thing happened in the afternoon. For the same reasons, if one touches the spadix of an Araceae flower as soon as it releases its revolting smell, one will observe that it is hot. If its temperature is measured, it might be as much as 10 °C higher than the air temperature. This high temperature helps release the amines and scatoles. How do plants manage to overheat on such occasions? After all, plants

Why do some flowers stink to high heaven?

How do some flowers deceive their pollinators?

Why do some flowers get hot?

are not warm-blooded organisms; they are at the mercy of external temperatures and their body temperature follows that of the air temperature (within 1 °C to 2 °C)

A biochemical energy valve activated at the right moment

The mechanism plants use to increase their temperature is biochemical, and its technical name is *thermogenic respiration.* As described elsewhere in this book, cellular respiration is a process of gradual and controlled oxidation (which is why it requires oxygen) of organic substances (mainly sugars), the end product of which is CO_2 (which also is released when humans exhale). In essence, respiration transforms the chemical energy contained in the sugars into chemical energy in the form of adenosine triphosphate (ATP), a very flexible molecule that can be used in almost all cellular energy transactions. ATP may be thought of as an energy currency: the particular value of every cellular substance (product) or process (service) may be reduced to ATP equivalents, just as products and services within a society are reduced to a common currency base. In even simpler terms, a citizen can sell a certain amount of sugar and use the money to buy a certain amount of meat. At the cellular level, through the respiration process, the cell oxidises a certain amount of sugar and gains a certain amount of ATP, which it then may use to manufacture an expensive product, such as lipids. However, as with all energy transformations, respiration is not a 100 % efficient process. Not all sugar energy is turned into ATP. One part, normally 35 %, is turned into thermal energy, which heats the respiring body. In other words, the efficacy of respiration is 65 %. Warm-blooded organisms are equipped with an internal thermostat that regulates the rate of respiration, so that their body temperature remains constant. When the interior temperature drops (e.g., because it is cold and a person is not dressed properly for this weather), the respiration rate increases so that more internal heat is produced. Of course, this also means more ATP is produced and more sugars consumed.

How can heat production increase without consuming many sugars and without producing extra ATP (like printing inflationary money)? The manner selected is the presence of a biochemical valve that regulates the energy flow in one direction or the other, in effect changing the relative efficiency of the production of ATP and heat. When the valve starts operating, the efficiency of ATP production falls from 65 % to 22 %, whereas that of heat generation rises from 35 % to 78 %. This capacity has been recorded among bees, which heat their hives the same way in winter but do not need high energy in the form of ATP, because their activity is restricted by low temperatures. A second case relates to newly born humans. When a mild increase in respiration is not sufficient to regulate an adult's temperature adequately, he or she involuntarily starts to shiver. Shivering is a fast muscular movement that consumes large amounts of ATP. Because this muscular movement is not translated into body motion (i.e., kinetic energy), the excess energy is turned into heat. The muscles of newborn babies cannot shiver, but they have the capacity of thermogenic respiration within the so-called brown fat cells, which are also manufactured by animals that hibernate. In effect, this mechanism is the same as that of plants, but it differs with regard to the biochemical details – among animals, the heat generation efficiency of respiration is closer to 100 %. It seems, therefore, that the trick is quite common, but its adaptive value varies in different organisms.

The valve is found in the flowers of certain plants and in bees, newborn humans, and hibernating animals

In the case of Araceae family plants, the spadix (the cylindric organ of the inflorescence) is full of starch. It also contains small quantities of amines and scatoles, sufficient to create the smell that is so revolting to humans. However, in the temperatures of spring, when the plant is in bloom, no odorous substances are emitted. When the male and female flowers mature, however, spadix cells receive a signal from the now-mature flowers to initiate thermogenic respiration, that is, to activate the valve that will direct the energy generated by sugar respiration into heat. The signal is chemical. What also is interesting is that the substance responsible for the signal is salicylic acid, an acetylated derivative of which (acetylsalicylic acid) is common aspirin. Before it became possible to chemically synthesise aspirin, some plants were used in its preparation. Some of these plants – such as the willow (*salix* in Latin, from which the name derives) – have a high concentration of this substance. Most frequently, however, it is found in trace quantities and seems to serve as a signal in two cases: from Araceae flowers to the spadix as a trigger for thermogenic respiration and from areas that have been attacked by pathogens to intact areas, so the latter may prepare for a potential attack. When the signal from the flowers reaches the spadix cells, the starch breaks down and the sugars produced enter the biochemical pathway of thermogenic respiration. The cost of this complex mechanism is significant. Depending on the type of plant, the cylindric spadix may be as small as a baby's pinkie (e.g., in Mediterranean Araceae plants) or as big as a sizeable bottle (e.g., in some tropical Araceae plants). This means the starch quantity invested in thermogenic respiration may be as great as several tens of grams. In other words, a significant percentage of the photosynthetic product is consumed to achieve cross-fertilisation and, therefore, diversity within the population. The cost of volatile amines also should be included in this expenditure, along with the costly polysaccharide-rich sticky substances used to temporarily entrap flies.

Why do plants synthesise aspirin?

How do some plants pierce ice?

Are there more registered plant cases of thermogenic respiration? What do they use it for? If one climbs a mountain on a spring day and reaches the zone above the trees, he or she may notice crocus flowers emerging from the ice. Many plants flower before they sprout leaves. Crocuses spend the winter in bulb form underground; the bulbs store the photosynthetic products the plants produced the previous summer. In early spring – sometimes even before the snow melts – they flower using these summer photosynthetic products as raw materials. How do crocuses manage to penetrate the ice and emerge? As in the Araceae plants, thermogenic respiration raises the flower's temperature and not only melts the ice but allows the flower to maintain intense metabolism rates. The role of thermogenic respiration does not end there, however. The flower is warmer than the air, even after emerging from the frozen ground. This is very important because, at high altitudes, insect activity is reduced in early spring because of low temperatures. These early plants therefore become thermal refuges for cold-blooded organisms within an inhospitable environment. These are places where insects of both sexes meet and mate. What other place could be better? The insects' reward for their pollination services is multifold. These flowers provide room and board as well as a love nest for a pittance.

The flower as a thermal refuge

The flower as an insect's love nest

Another tactic plants use to keep their flowers warm at high altitudes is heliotropism, a motion that allows the plant to follow the path of the sun during the day and maintain the sun's rays perpendicular to the flower surface. The signal for this movement to start is the blue region of the light spectrum. This motion is reversible; the plant will turn back to the east the following morning. The mechanism behind heliotropism is hormonal. The light causes an uneven distribution of a growth hormone and its corresponding membrane receptors on either side of the stem; this results in fluctuations, depending on the relative growth rates of the two sides. Using one (thermogenic respiration) or the other (heliotropism) method, these plants manage to maintain their temperature as high as 10 °C above that of the air.

Why are some flowers heliotropic?

Why do some flowers produce chemicals used by insects to approach individuals of the opposite sex?

The previous case involving flies and excreta odours is one of pure deception in which all normal terms of an honest business deal for the provision of services at a fee have been violated. There are many other forms of deception. As already stated, many insects produce volatile substances in special glands and release them into the environment. These substances (called pheromones) are characteristic of both species and sex. For example, the female pheromone indicates the presence of heat and provides a pathway for males to find mature females for reproduction. The huge biochemical arsenal of plants has found ways to produce such substances and use them deceptively to attract male insects to flowers. A typical case is that of some *Ophrys* orchids, which are pollinated with the help of wild Andrena bees. The flower produces and releases into the atmosphere δ-cadinene, a terpenoid, which is the pheromone of the female insect. The unfortunate male approaches the flower in a state of arousal looking forward to carnal pleasure with the female of its dreams. Of course, this olfactory swindle would be ineffective if it were only olfactory. However, instead of being able to swallow its disappointment and move on, the hapless male bee is in for a second act of this drama, which is even more ludicrous. The flower looks extraordinarily like a female Andrena: perfect antennae, ideal eyes, exquisite curves, excellent abdominal lineation, and a perfume to drive the male bee crazy. In total abandonment of consciousness and self-control, the bee mounts the flower and "couples" with it. In its vain attempt to penetrate, the bee shakes the orchid's stamens, and before it realizes the ridiculous situation it has got itself into, a pollen shower has sprinkled his body. Incurably aroused and uncontrollable it will seek new adventures. If it finds a female bee, it will propagate its species. If it ends up with another flower, it will contribute towards the propagation of another species. The plant has committed a perfect crime. In biological terms, it has

To serve their own interests, plants often usurp the reproductive heat of insects: a ludicrous trick

Why does the Andrena bee try to copulate with the Ophrys orchid flower?

increased the likelihood of its genes contributing to the new generation, by conning the insect not only through olfactory but also visual deception, and without paying any carriage fees for the pollen transported.

Trick and Truth

By now, it is apparent that the species of a plant's pollinator often can be predicted by taking into account the shape, size, colour, and position of its various flower parts. However, the relationship between plants and their pollinators is rather loose; in most cases, there is no exclusivity involved – all that matters is being in the right place at the right time. Therefore, a plant may exploit more or fewer insect visitors and every insect, similarly, may visit more or fewer plants. The prerequisite condition is that plant and insect match with regard to morphology and physiology. Furthermore, the flower fixation of a bee lasts as long as the blooming period of its favourite plant; once that ends, bees turn their attention elsewhere. Still, there is a relationship that is absolutely loyal and exclusive and would make Odysseus' Penelope or any other symbol of spousal faith pale before it. The relationship between plants of the *Ficus* genus (including the well-known fig tree, *Ficus carica*) and some tiny wasps has been forged through geologic aeons and probably has remained unchanged for tens of millions of years.

A cohabitation of deep loyalty, or why a fig tree cannot divorce the wasp

There are about 750 *Ficus* species, most of which thrive in the tropics and hot climates, with very few species encountered in the temperate zones. In most cases, each species has its own unique pollinator, whose life is so closely related to the plant that the insect spends only a tiny part of its life outside the plant. Most of the time, the insect lives within the plant, within a flower, which becomes its nest.

The "fruit" of the fig tree is actually a closed inflorescence

Like most plants, fig trees have flowers, but these are hidden and seen only when one opens a fig. Contrary to popular belief, the fig is not a fruit, it is a flower cluster, a composite flower head, that is, an inflorescence. As already stated, flowers ought to be apparent, to advertise their presence in a visual as well as an olfactory manner to attract pollinators. In a few plant species, however, some of the flowers that are to be self-pollinated do not need to be conspicuous and impressive. In the case of the fig tree, the necessary, tough exclusivity has doomed its flowers to total obscurity. Like a jealous spouse, the fig tree hides its flowers so they can be found only by an initiated insect. Therefore, what is erroneously called a fruit is actually a closed composite flower – a collection of several tiny flowers – fixed in the fig's inner surface within its tasty red jelly, and it is totally obscure. The tiny entrance to the interior is at the tip of the fig, known as the ostiole, and it is quite impressive. The ostiole comprises successive scale-like layers that can be pushed inwards but not outwards; in other words, whoever enters remains trapped inside, and should have a very important reason for being there.

Actually, the close bond between the plants of the *Ficus* genus and the wasps that pollinate them presents all the characteristics of a strict symbiosis: Neither of the two partners can live without the other, or, to be more specific, neither can reproduce without the other. If one of the two disappears, the other one has no future. The weakest link is the wasp, which only lives for a few days, while *Ficus* plants may live for tens or hundreds of years.

Why do figs contain sterile ovaries?

There are many variations in the details of this symbiosis. However, one of the simplest does not concern *Ficus carica*, the cultivated fig tree. In certain species, the closed inflorescence contains three types of florets: fertile female, sterile female, and male florets, each with its own function. Fertilized female wasps no longer than 2 mm (i.e., the size of a pinhead) are attracted by scent to the mature figs and enter their interior through the ostiole. However, it is not easy to enter. The scales of the entrance (known as bracts) resist, and the wasp has to push and edge its way through. It does not give up, however, because it has every reason to enter, even if it becomes crippled by the effort. Indeed, the wasp usually ends up without antennae or wings. This is of no consequence, however, because the wasp is destined to lay its eggs, pollinate the female flowers with others' pollen, and die. Even if the wasp wanted to come out, it could not. The scales at the entrance cover one another in a manner that allows the wasp to push its way in, but to exit, the scales must be pulled in one by one, which is very inconvenient for the wasp, having only two front legs. On the other hand, even if the female wasp managed to exit, it would not go very far; it cannot smell (i.e., direct its flight) without antennae and no longer has wings to fly.

The tiny wasp enters the fig to lay its eggs in the sterile ovaries, to pollinate the fertile ones, and to die

Where did the female wasp find the pollen from other plants? How does the mother behave, and what do her offspring do when they hatch? Unlike humans, who value their privacy, wasps remain indifferent if one opens a hole in the fig large enough to observe what is happening inside the inflorescence. The wasp's behaviour and movements are instinctive and uninhibited, even if the intrusion is particularly indiscreet. Scientists, who are habitual voyeurs, can make canny and usually legitimate observations of animal behavior without running the risk of being labeled *peeping Toms.* Perusing relevant reference literature, one can read scandalous descriptions of the coupling of mussels, grasshoppers, donkeys, and humans using extremely scientific and precise terminology. Therefore, it has been discovered that the female wasp lays its eggs on sterile female flowers, the stylus (i.e., the tube from the stigma to the ovary) of which is the same length as the egg-laying apparatus of the wasp. The wasp, in other words, lays its eggs on the ovaries that are sterile and

How can the wasp differentiate between the two types of ovaries in fig tree flowers?

will never produce seeds. At the same time, the wasp takes two more initiatives: When laying eggs, the wasp leaves a drop containing substances that stimulate growth of the sterile ovary, thus creating a bulge of plant tissue so that the hatching larvae may find a source of food. The wasp also uses its front legs to empty the pollen-carrying sac within its thorax (similar to that in the hind legs of bees) and scatters the pollen on the stigmata of the fertile female flowers. These flowers have a stylus that is longer than the wasp's egg-laying apparatus; the wasp has learned to distinguish between the sterile and fertile flowers and attempts to lay its eggs on the former while scattering the pollen on the latter, having checked the length of their stylus. This process prevents the wasps' larvae from developing where the plant's seeds are located. When the wasp has laid all the eggs and scatters all the pollen grains, it has fulfilled the aim of its life and dies.

The fig seeds and young wasps grow side by side within the same fig

Within this enclosed, protected space, the offspring of two totally different organisms grow side by side, because nature has entangled them in an eternally loyal relationship. The plant sacrifices part of its ovaries to another organism's development for the sake of having its pollen transported to achieve cross-fertilisation.

The miserable life of the male wasp, which never sees the sunlight

The great metabolic activity within the closed area, where the wasp's and the plant's offspring develop, results in an increase in CO_2 and a reduction in O_2 concentration. Under these conditions, young male insects are active whereas the females are dormant. Therefore, the males fertilise the females on the spot while the latter are asleep and show no resistance. It might be better this way, because if the females were awake, they might never accept such wingless, blind, and sickly creatures that perform only two tasks during their brief (a few hours long) life: they copulate with a sleeping female and open a hole in the wall, right next to the ostiole. At that point, their aspirations and life's purpose have been fulfilled and they die. Not only do the males of this wasp species live for a very short time, they also never see sunlight. As far as propagating their packet of genes, though, they do exactly what is necessary. The hole in the wall of the composite flower (i.e., the fig) restores the balance of the interior and exterior atmospheres, CO_2 levels decrease, oxygen increases, and the females wake up and realize that conception has taken place, without even knowing which of the males, whose bodies are lying around, was the perpetrator. Having nothing else to do, the females escape through the hole the males opened, having filled their thoracic sacs with pollen from the male flowers near the hole (remember that the natural fig opening – the ostiole – cannot open from the inside). A characteristic feature is that male flowers mature and produce pollen only during the final phase of the drama, shortly before the females awaken. In contrast, female flowers are mature from the very beginning, waiting for the wasp-mother to pollinate them with pollen from another fig individual. This is how unwanted self-fertilisation is avoided. As soon as the final female wasp escapes and the plant's seeds have matured, the inflorescence (fig) changes colour and becomes softer and sweet, that is, it becomes a ripe fruit, ready to be eaten by animals so that its seeds may be dispersed.

In search of a new inflorescence

What about the female wasp? What does it do once it has escaped? Using its sense of smell, the wasp searches for an intact fig at the phase of the female flowers' maturation, so as to conclude its life cycle. Indeed, the female wasp must hurry because it has only 2 to 3 days left to live, and it may take that long for the wasp to find an appropriate fig that will provide it with room, board, and protection for its offspring in exchange for the load of pollen it carefully carries in its thorax.

The wasp's size and flying capacity does not allow it to travel far. Therefore, the wasp must quickly find a fig in the first stage of its development, with mature female flowers, where the insect may lay its eggs before it dies.

The easiest solution would be for the wasp to inhabit a fig on the tree where it was born. However, this would be impractical because the maturation of all the flowers of an individual is synchronised. In other words, all the figs on the tree where the wasp spent the first hours of its life and had sex (in its sleep) are in the same phase; their fruit is ready to fall. Therefore, the insect has to find another plant. In this manner, of course, the plant has ensured cross-fertilisation, even if the insect goes to some more trouble. Nature, however, has foreseen this difficulty. The populations of *Ficus* species are so dense and the distance between them so short that the wasp has enough time to find another tree – otherwise, the wasp would disappear first, followed, some centuries later, by extinction of the *Ficus*. There is a prerequisite, however: there should always be some individuals in the population with inflorescences (figs) in the right phase. In tropical regions, where most *Ficus* species thrive, individuals flower in succession, so there are always some fig trees suitable for egg-laying. Therefore, while some individuals have mature fruit from which the female wasps emerge, other individuals nearby have maturing flowers ready to receive those female wasps. Ultimately, the flowers of each plant mature simultaneously, whereas the flowers of successive plants mature in succession.

Why should tropical fig tree populations be dense?

How can symbiosis be secured when fig tree populations are thinned?

What if the neighbouring individuals are lost, if the density of the population is reduced? Obviously, such disasters have occurred in the past. How did the wasp survive? It would not be a surprise if an effective solution were found for this problem as well. Certain individuals within the fig population disobey the rule and do not coordinate the maturation of their flowers. They keep producing flowers at various maturation phases so that the wasp exiting a plant need not go far. This works well for the wasp, but it practically obliges the plant to accept self-pollination. In other words, some individuals in the population sacrifice the privileges of cross-fertilisation to ensure the survival of the wasps scattering pollen around the population. These individuals are the centres from which wasps will spread out in case something goes wrong. It is the wasp's refuge in difficult times.

What about *Ficus* plants that live in more temperate climates, where there are significant seasonal fluctuations in temperature and humidity with corresponding seasonality in insect activity? As cold-blooded animals, insects have difficulty flying and moving during the winter. Many fig trees in these regions produce sterile flowers (figs) in winter; these flowers are virtually dormant and are offered to the wasps merely as a wintering abode. These trees produce regular flowers in the summer.

What about the *Ficus carica*, the tamed fig trees of the Mediterranean? After the aforementioned discussion, the reader may wonder why when a mouth-watering fig is opened, it is not full of dead insect bodies. Remember that the females that enter to lay their eggs do not exit; they die inside. The same is true of the young males hatched from the eggs. They copulate, help the female escape, and die inside the fig as well. Therefore, one might expect a fig to be a wasp cemetery, but this is not so. It should be underscored at this point that humans find insects revolting to eat, at least in regions where there is no scarcity of proteins. Our fellow humans in sub-Saharan Africa, where meat and milk are scarce, make exactly the opposite choice, however: they select fruit with worms in an effort to supplement their diet with protein. If we are given a choice of a fruit with a worm and a fruit sprayed with insecticides, we would choose the latter.

The fig tree is one of the earliest "tamed" plants

The cultivated fig tree was "tamed" at least 7,000 years ago in the so-called Fertile Crescent (present-day Iraq and its surrounding region) soon after wheat and barley had been tamed. As is true of all other plants, humans selected from wild populations the fig species with the most desirable features, in this case, figs without dead insect bodies. Today, there are several varieties of cultivated fig trees. One variety is a mutation of the wild type and needs no pollination by insects. It is parthenocarpic, that is, it produces fruit without pollination but produces no fertile seeds. In other words, it can reproduce vegetatively only through cloning. The only requirement is a branch of appropriate length and thickness, which undergoes the right processing so it may produce a new plant genetically identical to the original, that is, sterile and with no need for pollination, which means no dead insects in the fig. The seeds sometimes observed in overripe fresh figs or in dried figs are sterile. These varieties are direct descendants of those chosen to be cultivated by ancient Semites and have been preserved by humans through the centuries. Other varieties have come about as the result of mutations that do not produce infertile female florets. In these cases, the female wasp enters the inflorescence but does not find suitable ovaries on which to lay its eggs. Through its effort, though, it lays the pollen on the fertile florets and pollinates them. The wasp then dies without any gain, and her body decomposes as a result of the enzymes secreted by the inflorescence. Therefore, these figs also have no insects.

Why are there no wasps in cultivated figs?

Loyal wasps and con-artist wasps

However, someone finding a tiny worm (i.e., an insect larva) within a fig might wonder how it got there, as wasps do not enter the inflorescence. The answer is, the worm was found there because there are impostors that refuse to pay rent. Pollinator wasps go through the main entrance and use the fig as accommodation, paying for it by laying pollen. However, other, parasitic wasps also exist that lay their eggs by making a hole in the fig wall, without entering or making contact with the stamens. These wasps do not carry pollen nor do they assist pollination. They simply use the fig as a place to stay, eat, and shelter, without paying any of the known rewards.

Why do Plants Manufacture Costly Fruits?

Dormancy and dispersal

The next important event in plant reproduction and geographic spread is seed dispersal. When ovary fertilisation concludes successfully, the flower parts that have completed their mission age and drop. The remaining parts are involved in the process of creating a fruit that contains the seeds. The embryo matures, and the seed, to a greater or lesser extent, is supplied with stored nutrients that will support the growth of the new shoot until it starts photosynthesising and absorbing water and nutrients through its roots. At the same time, the ovary walls and, in some cases, other flower parts, such as the sepals and anthers, form the fruit that contains the seeds.

The fruit is an invention with a double function: to prevent the immediate sprouting of the seed and to facilitate the seed's dispersal. In the wrong season, it would be a failure for a seed to sprout immediately after it falls from the mother plant. Mediterranean annual plants, for instance, usually bloom in the spring, when the weather conditions (temperature and wet soil) warrant high photosynthetic rates that can afford the cost of flowering and fruit ripening. The fruit is released at the end of spring or the beginning of summer. If the seeds sprouted immediately, the young plants would have to face the long, mercilessly dry Mediterranean summer ahead. As for more temperate environments, where plants flower during a wet and warm summer and release their fruit in the autumn, if seeds sprouted immediately, the young plants would have to face a frosty winter. Both cases represent a tragic reproductive failure.

Why are seeds dormant when they depart from the mother plant?

For this very reason, seeds need to be equipped with temporary mechanisms that inhibit sprouting; in other words, they must be given the opportunity to remain dormant. This is what happens during seed maturation and fruit ripening, while the seeds and fruit are still on the plant. There are two types of dormancy: mechanical and hormonal. In the former, the outer covering of the seed or fruit turns into wood, becoming impermeable to water and oxygen. No seed can germinate in the absence of water and oxygen. An example of a fruit with hard seeds is the olive. The flesh of the fruit is soft, but its stone, which contains the seed, cannot be opened, even with the use of pincers. Almonds, on the other hand, have a hard fruit shell, whereas the seed is relatively soft. Obviously, mechanical dormancy ends when the hard, wood-like outer cover is eroded in some way. Until then, however, a critical period has passed, and if the erosion is delayed sufficiently and completed at a time suitable for sprouting, then the likelihood of the young plant's survival increases. How is erosion achieved? In the worst-case scenario, the fruit is left at the mercy of natural phenomena. Frost and thaw, friction against the hard ground, attacks by microorganisms, or even fire will eventually create a hole through which water and oxygen can enter and the rootlet of the young plant can exit.

Why are the outer covers of fruit and seeds so hard in some plant species?

How is mechanical dormancy terminated?

The case of fire is of particular interest in ecosystems (e.g., the Mediterranean ones) in which frequent spontaneous fires are an inherent natural feature. The accumulation of large quantities of biomass and its fast drying during the arid and warm summer make Mediterranean ecosystems combustible and vulnerable to fire. The long coexistence of plants and fire has led to typical adaptations of some plants, so not only is their survival not threatened by frequent fires, but their spread actually increases after such fires. The outer cover of seeds in Cistaceae family plants (rock roses) does not allow sprouting, unless the fruit has been roasted for some hours at a high temperature. These plants produce many seeds every year (an average bush may produce several thousand seeds); these seeds remain dormant for decades in the soil seed bank. They are activated and sprout following a fire and inundate the post-fire landscape as if they were a

In some Mediterranean plants, dormancy is terminated when seeds are roasted by fire

monoculture. Such seeds, along with archaeobacteria at thermal springs, are probably the only biological material that can tolerate temperatures near 100 °C for several hours. Moreover, the dormancy is terminated at the most appropriate time, given that the landscape has been cleared by the fire and competition from other plant species is minimal.

In many fleshy and edible fruits with hard stones, the termination of dormancy requires that the seeds pass through the digestive tract of animals. The gastric fluids erode the hard walls, so after defecation of the stones, the seed can sprout. The system is noteworthy in that it is multifunctional and of considerable adaptive significance. The fleshy fruit usually has bright colours advertising its ripeness and high nutritional value. Before it ripens, the fruit is usually green, hard, and sour. Its colour indicates it is not suitable for consumption, and nobody attempts to eat it; this benefits both the plant and the consumer. Unripe fruit is unpleasant to the palate and indigestible. As for the plant, the visual warning about nonripeness saves the seeds, which are not yet fully developed, and their outer covers are not hard enough. When the seeds have developed fully, hormonal signals send a mandate for drastic biochemical changes in the fleshy part so that it becomes edible. The aim is not only to attract customers, but for customers to try it, find it pleasant, and devour the fruit and the seed contained within. The hormonal signal, produced inside the now-mature seeds, is a very simple organic substance called *ethylene*. Ethylene activates the genes that produce the proteins that act as catalysts for the partial breakdown of the cellular walls of the fruit flesh. This is how the fruit softens. Other primed genes activate the biochemical path for the production of pigments, similar to those that colour flowers. Finally, the acids that made the unripe fruit sour now turn to sweet-tasting sugars. Sometimes, volatile, pleasantly scented substances also are produced.

Why do some plants produce fleshy fruit with hard stones?

Fruit ripening: a commercial message for fruit-eating animals

The whole operation, therefore, aims to protect the fruit while the seed is being prepared and to advertise it (through colours, scents, and flavours) when the seed has matured. When an animal swallows a seed, there is a fourfold value for the plant. First, the potent gastric acids wear down the hard perisperm to a certain extent, facilitating the effects of subsequent natural erosion. Second, animals do not defecate where they eat; rather, an animal will dispose of the useless load some distance from its food source, contributing to seed dispersal. Third, the seed will be placed in excrement, an environment teeming with nutrients. Fourth, any soft seeds from other plants the animal may have consumed have already been destroyed by gastric fluids, meaning the harder seed has been placed in a rich and nutritious environment where there is little competition. Obviously, the case just described has many things in common with animal-assisted pollination. The plant packages its seeds attractively, advertises its fruit with colours and scents, and supplies the packet with highly nutritious content. The animal receives its reward and in exchange inadvertently transports the seed to a spot suitable for sprouting. If the fruit is not eaten, it will rot under the influence of microorganisms and the coat of the released seed will be eroded only with the help of natural erosive factors, not having benefited from the advantages of passing through an animal's digestive tract. From this point of view, the habit humans have of spitting stones and defecating in toilets is unfair and improper if one considers the trouble the plants go through to produce the fruit.

It is truly worth producing fleshy fruit: there are four reasons why it is worth having a stone pass through an animal's digestive tract

Fleshy fruit: a new case of honest transaction

The hormonal method of imposing dormancy is based on the seed's being equipped with the right growth inhibitors. In this case, very hard outer coats are not really necessary and the seed may be steeped in or absorb oxygen, but the presence of the inhibitor prevents the onset of sprouting. The hormone that plays this role is abscisic acid, the same hormone that functions as

When seed dormancy is imposed hormonally

a signal in another case as well. When there is no water in the soil, abscisic acid is produced in the root and transported to the leaves, where it induces the closure of stomata to avoid the threat of dehydration. In fruit, this hormone accumulates in the seeds following their maturation. Hormonal dormancy terminates with the gradual reduction of abscisic acid concentration, which may occur spontaneously or be accelerated by environmental factors. For instance, abscisic acid is difficult to dissolve in water but is not completely insoluble. When seeds are rinsed with water, some of the hormone is removed. When the concentration is reduced below a certain level, seed sprouting is no longer inhibited. In other words, in natural conditions these seeds operate as rain height sensors, or pluviometers. When the rainfall exceeds a certain level (i.e., when the hormone has been reduced to the "right" level), the seed sprouts. This mechanism ensures that a little rain will not trick the seed into sprouting prematurely and allows sprouting to start only when sufficient supplies of water are guaranteed. This system may be effective in Mediterranean plants so that they do not sprout after an odd summer rain but only after frequent autumn rainfalls. Other seeds require a period of low temperatures – usually 2 to 4 weeks – before hormonal concentration levels drop below the threshold and give the signal for sprouting to start. This phenomenon is called *vernalisation*, and its adaptive significance is that seeds will not sprout until heavy winter has passed.

Seeds with rain sensors

Seeds with cold sensors

Plant Migration Takes Place at the Embryonic Stage

All this leads to the conclusion that seeds require not only dormancy mechanisms but dispersal mechanisms as well. It would be a failure for a seed to sprout right

There is no parental care among plants: offspring must hit the road

under the mother plant, because it would have to face competition with the latter for the exploitation of local resources. This is a case of interspecies competition, that is, competition between two individuals of the same species with similar exploitation potential but of different ages. The young plant cannot compete against its mother, which has already been subjected to the pressures exerted by the particular environment and has had ample time to adapt. If the young one establishes itself further away, it might have a better chance of surviving, even if it is crowded by plants of different species, that is, plants that apply slightly different ways of exploiting resources. There might be some free ecologic niche somewhere for the young plant to develop its own adaptive mechanisms. However, dispersal mainly satisfies a species' need to expand its vital territory. For these reasons, natural selection has favoured the development of multiple characteristics integrated in the seeds and fruit to enhance their dispersal. It was already mentioned that the production of fleshy fruit serves exactly this purpose, that is, the mutually beneficial relationship between animal and plant, similar to the relationship that serves pollination. Animals, however, also may contribute to dispersal passively. Seeds or fruit with a sticky surface or hooks may cling to the fur of passing animals or the feathers of birds and may be carried over long distances. Velcro, the well-known adhesive in strips with its tiny hooks and many applications, was invented by a Swiss mountain climber who was curious enough to observe under a stereoscope the surface of the bothersome fruit that stuck obstinately to his trousers. Also, one should not forget the passive transport by animals, without any suitable adaptation on the plant's part. For example, if one walks in mud, scrapes it off his or her boots, and places it in a flowerpot, he or she would soon see that walking is good not only for the heart, but for plant dispersal as well.

Seed dispersal is the only way for plants to migrate

Passive migration via animal carriers: why do some seeds have hooks?

Parachute-like seeds and helicopter-like seeds

Other plants produce fruit or seeds that are light and equipped with mechanisms for hovering in mid-air. The parachute- or umbrella-like seeds of dandelions with their tufts of silky hair or the wings of the helicopter-like maple tree seeds are typical examples. Seeds are carried not only by the wind but also by water, provided they float; this is how coconuts colonised tropical islands. Finally, some plants actively undertake their own seed dispersal. The green fruit of the squirting cucumber plant (*Ecballium elaterium*) has elastic walls. The fruit swells so much with water that its internal pressure reaches almost 10 atm, an impressive magnitude considering the pressure in a car tyre is two to three times less. The weak point in the fruit is where it joins the stem. When the internal pressure exceeds the resistance strength, an almost imperceptible touch or a strong wind is enough to cause the fruit wall to tear off at this weak point, scattering its seeds over a distance of several metres. The spatial arrangement of stem and fruit is such that the point of seed escape either faces upward or is at a small angle with the vertical plane. The seeds, therefore, are ejected according to all ballistic laws, and the fruit operates like a catapult.

Catapulting fruit

Plants produce many seeds to compensate for their high infant mortality

Why do plants produce so many seeds? It was mentioned earlier that a medium-sized, 50-cm tall rock rose bush produces a few thousand seeds a year. Most of them fall victim to specialised fruit- and seed-eating animals that manage to circumvent mechanical and chemical defences or fall prey to an attack by pathogens and rot. Those that escape from their predators as seeds and sprout are very vulnerable as young plants, with soft and easily digestible shoots. Furthermore, young plants are sensitive to abiotic stress factors, mainly the lack of water. All these parameters result in a high mortality rate. If 1,000 lentil seeds are planted in a sterile and controlled space, it is possible that 900 may sprout. If the same number were sown in a field, perhaps fewer than 10 would sprout.

A summary of the plant's reproductive system and related functions

The conclusion is that the problems faced by plants due to their sedentary nature have been solved through the adoption of a flexible system including vegetative reproduction by cloning, the possibility of self-fertilisation, and the use of several tricks to transport gene packets from one plant to another to achieve sexual reproduction, which is ecologically preferable. The pollen carriers may be physical (wind) or biological (animals). In the latter case, relationships develop between plants and their pollinators that are based on rewards provided by plants in exchange for the reproductive services provided to them by the animals. In this case, the parties involved present a range of morphologic and physiologic convergence patterns, characterised by target aiming. The body of the insect and the shape and size of a flower should match. Furthermore, the insects' visual and olfactory perceptive capacities must be compatible with the visual and olfactory signals emitted by the plants' flowers. Finally, the reward provided ought to be compatible with the nutritional demands of the animal. The coevolution of the flower and its pollinators through geologic aeons has led to an increased level of loyalty in their mutual relationships, so sometimes the survival of one partner depends on the existence of the other. This conclusion is more significant for the animals, however, because plants can turn to cloning through self-pollination or vegetative reproduction if their partner disappears.

In some cases, the honesty of the relationship is violated and there are phenomena of deceit. When fertilisation has been achieved, fruit production begins. The fruit ripens at the same time the seeds it contains mature, whereas temporary hormonal or mechanical dormancy is imposed on the seeds. The fruit protects the seeds and serves their dispersal, which is assisted by the wind, water, and animals, and fruit is suitably adapted to serve this purpose. Fleshy fruits are aimed at consumers; they advertise themselves suitably to appeal to animals, which they provide with food in exchange for seed dispersal. Seed dispersal is the only way plants can expand their vital territory or migrate to more suitable areas when the climate changes. Animal-assisted dispersal is particularly efficient and covers a wide radius. It is particularly significant during climatic changes, which cause active migration of animals and passive co-migration of plants in the form of fruit and seeds. The new territory is selected by animals, but its climate will suit plants because it suits animals, given both organisms originated from the same climate. For instance, in the past two million years, Europe has gone through repeated major and minor ice ages, during which regressive migrations of both animals and plants have occurred. Paleontologic records indicate that plants, despite being sedentary, did not lag behind in migration speed. Therefore, seed-dispersal tactics are as effective as running.

Chapter 6
The World through the Eyes of Plants

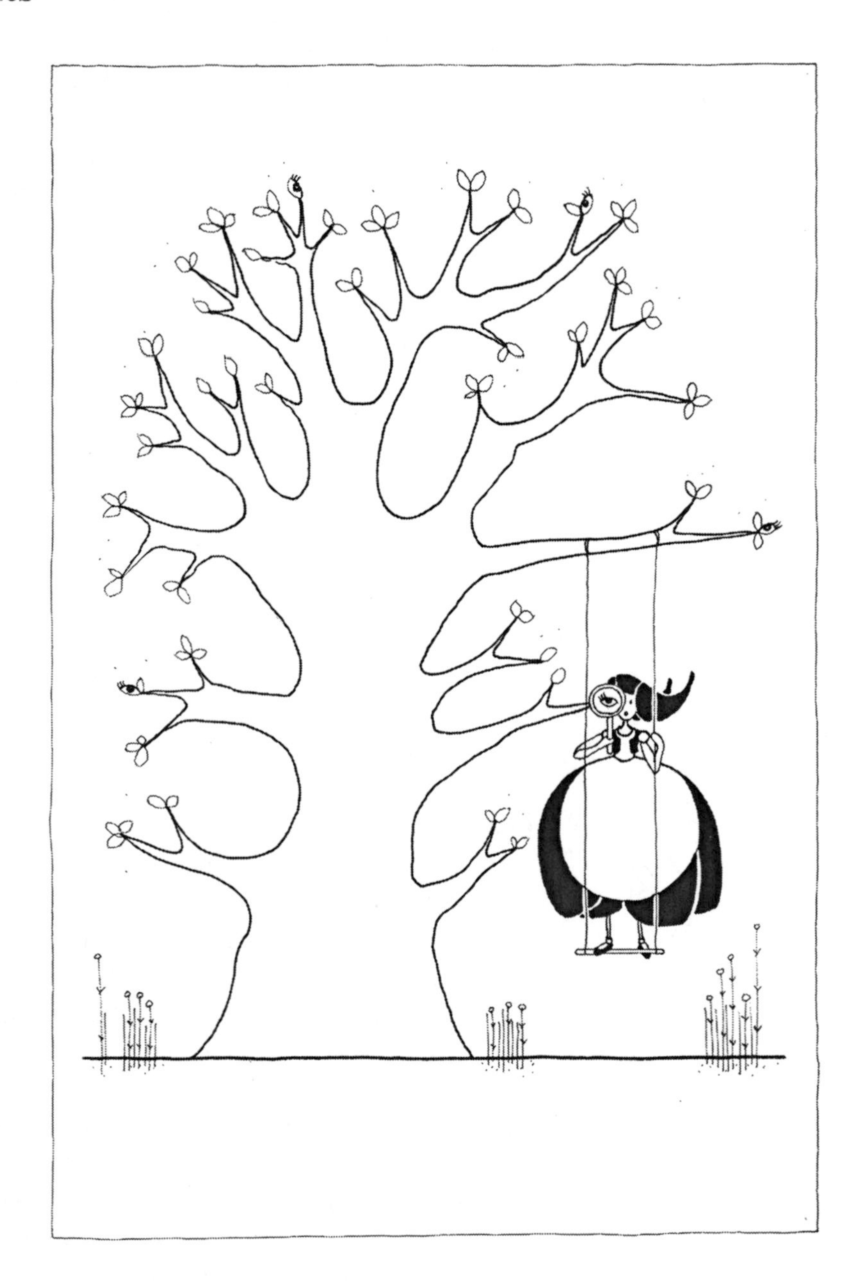

What and how do plants feel?

The notion that plants might be capable of "feeling" used to have ardent supporters and fanatic opponents. Around 30 years ago, a book was published in America titled *The Secret Life of Plants*. There have been many editions and the book was translated into many languages, including Greek. Justifiably, the Greek translation was published as part of a parapsychology series, as the book was not written by biologists and was aimed at impressing rather than informing the public. What the book did was attribute to plants feelings of love, hatred, rage, sorrow, pleasure, gratitude, joy, sadness, cheerfulness, moodiness, hope, surprise, and many more. In addition, the whole book was permeated by the view that not only are these emotions measurable, but that they had already been measured by scientists. Of course, the authors' argumentation could not stand up to serious criticism. The publishing success of the book, however, led two acclaimed biologists to send a letter to the editor of a prestigious scientific journal in an effort to restore the truth. *The Secret Life of Plants* is a mixture of real and imaginary aspects of plant behaviour, which its authors interpret in an inadmissibly anthropocentric manner. For example, they claim that the successful growth and good health of a plant well cared for by its owner create feelings of gratitude and love in the plant for its master. If its master suffers, not only does the plant feel discomfort, it also expresses it (in the authors' view) through electric or other waves that run through the plant's body and are measurable. Although acclaimed scientists supported the idea (and despite the ridiculousness of the matter, were willing to prove experimentally) that the intensity of the alleged "waves" was lower than that of the "noise" (i.e., the normal interference) of the measuring devices, the views expressed in this book enjoyed great popularity among the general public. In the United States, in particular, there was a battle to stop plants from being considered credible witnesses at

The need to shift from an anthropocentric to a phytocentric view of things

Unacceptable hyperbole and dangerous nonsense: plants as murder witnesses

murder trials or from being investigated. The rationale of those supporting such views was that if someone were murdered in front of his or her favourite plant, the plant would react upon viewing the murderer. Imagine an investigation with a lineup of suspects paraded before a wired-up plant that would identify the murderer in its own special way. Fortunately, common sense prevailed and the issue was forgotten.

During my first years of teaching at the university, I was amazed by the number of students raising questions in class regarding the truth of that book's claims and I was taken aback by some students' suspicions of my attempts to refute those claims with scientific evidence. Could it be that superstition is deeply rooted, or is there something fundamentally wrong with the way we perceive plant behaviour? Why are people (even students in their third year of biology) so willing to believe such weird things? At this halfway point in *Alice in the Land of Plants*, it should not be necessary to remind readers that the best way to comprehend the functions and behaviour of plants is to forget for a moment their own functions and behaviour as well as those of animal organisms more familiar to them. A dog might wag its tail in gratitude when offered food, but a favourite plant probably remains completely indifferent, from an emotional point of view, when it is watered. Loving and treating a fragile pansy tenderly might suit the character of some humans, but in no way does that mean humans should expect similar behaviour from the little pansy. What is natural for humans is supernatural and unfit for plants. In other words, what is needed is an intellectual shift from the world of animals to that of plants. One must stand in plants' "shoes" to restore the truth of the matter, to the extent our current scientific knowledge allows.

Why are people so gullible and superstitious?

Let us put ourselves in the "shoes" of plants and imagine what they would have to sense on the basis of their vital interests

In the Beginning, there was Light

Sense is the perception of stimuli

Basic concepts about the nature of light and the way it is perceived by organisms

The dual nature of light: wave and particle

A *sensation* is defined as the perception of stimuli. To clarify the definition further, one might say that through their senses, organisms perceive environmental stimuli. Light, for example, stimulates the eye. Light is characterised by intensity (strong or weak) and quality (colour, i.e., radiation wavelength). The sensory organ records intensity and quality, as well as their changes over time. All three variables comprise information about the light environment. Changes in intensity are particularly important (from strong to weak and vice versa), because they provide information about the movement of the light-emitting or -reflecting object away or towards the observer. The light that reaches the sensory organ originates from light sources (the sun, stars, artificial light sources). The rest of the objects do not radiate light but reflect light coming from the light sources. In other words, when light falls on a material surface, it might be reflected and change direction (an extreme example is light falling on a mirror's surface, which reflects almost all the light it receives) or it might interact with the material, that is, it might be absorbed. Any light not reflected or absorbed passes through the material. Another extreme case is that of transparent glasses or water, through which almost all nonreflected light passes. In other words, to comprehend the interaction between light and matter, one must accept that light behaves as a particle, that is, it has an impact on the molecules of the materials on which it falls. At the same time, light may be transmitted through space as a wave, with a specific wavelength (the distance between two successive peaks) and frequency (the number of peaks going through a point in the time unit). This dual nature of light (i.e., particle and wave at the same time) is expressed through the concept of *photons* or *quanta*. Photons are both particles (that collide with matter) and "packets" of electromagnetic energy transmitted as a wave. On impact, the energy of a photon may be transferred (under certain prerequisite conditions) to the matter, whereupon the photon ceases to exist. One then may say that the matter absorbed the light. Obviously, in this case, the matter has gained the photon energy and therefore its energy content has risen (think of the multitude of solar water heaters and photovoltaic systems – they all operate on the basis of radiation absorption by matter).

Visible (light) and invisible radiation

To be exact, stars and lamp filaments are not the only objects that emit radiation. In fact, all bodies, even if not lit, emit electromagnetic radiation in which the wavelength is inversely proportional to the body's temperature. In mild temperatures, the wavelength (λ) lies in the infrared region of the spectrum (> 700 nm, $1 \text{ nm} = 1/10^9 \text{ m}$), which does not stimulate the human eye; therefore, this radiation is invisible. At high temperatures (such as those on the sun's surface or the heated filament of a lamp), the radiation shifts to shorter wavelengths that include the so-called visible part of the spectrum (400–700 nm). This radiation stimulates the human eye and is perceptible. At even higher temperatures, the radiation shifts further to even shorter wavelengths, once again lying beyond the sensitivity range of the human eye. Although this radiation is also invisible, it is much more dangerous because it comprises high-energy photons (the energy of a photon is inversely proportional to its wavelength) that, when interacting with bio-macromolecules, oxidize or harm them. Examples of this are ultraviolet radiation and x-rays. Visible radiation, therefore, is that part of the natural electromagnetic radiation spectrum that is visible to the human eye. Other organisms are capable of detecting areas lying more to the right (infrared) or to the left (ultraviolet) of visible (to the human eye) light, but seldom beyond 750 nm in the infrared or 350 nm in the ultraviolet region. In other words, from the much

Not all organisms see the same radiation spectral region

Harmful radiation: when quantum energy is so high that organisms cannot manage it

wider spectrum of photons reaching the surface of the earth from natural sources (around 280–3,000 nm), the region from 350 nm to 750 nm has been selected by the organisms for use as information about their light environment. This is not a coincidence. Photons in the ultraviolet region, particularly those in the so-called UVB part of the spectrum (280–320 nm), are of high enough energy to be harmful. They usually cause ionization, that is, violent detachment of electrons from atoms, thus leading to oxidation. This harmful effect on the DNA is expressed as a mutation. Consequently, not only can this radiation not be used as information, but organisms must have protective mechanisms to block it from penetrating their cells or to repair any damage it might cause. This issue is discussed again elsewhere in this book.

Ultraviolet, visible, infrared: harm, information, heat

At the other end of the spectrum, the deep infrared region, photons have very low energy; therefore, they cannot cause the molecular reactions necessary to initiate and successfully conclude the process of light perception. Whereas the reaction of matter to natural ultraviolet radiation is violent and harmful to molecules and organisms, the effect of natural infrared radiation (λ = 700–3,000 nm) is so mild that, if anything, it might only warm organisms, particularly in the wavelength region absorbed by water.

In the part of the spectrum visible to organisms, however, the photon energy is exactly sufficient to meet relevant needs. What are these needs, then?

Light in sensory organs is absorbed by molecular photoreceptors

Animal organisms are equipped with light-receptor molecules capable of absorbing light. Usually, these light receptors are concentrated in sensory organs, that is, the eyes. Light absorption results in a change in the molecular structure of the receptor; this change is not permanent, but transient and reversible. It may be expressed in a simple formula:

$$A^+ \xrightarrow{\text{light}} A^-,$$

where A^+ and A^- are the two molecular forms that differ slightly from each other, but enough to generate a signal. The more light there is, the more the ratio of the two forms of the compound shifts to the benefit of A^-, and vice versa. The A^+:A^- ratio is transformed into an electric signal transmitted through the nerves to the brain, where it is perceived by the organism as information about the light environment. A low-intensity signal might mean that the whole of A is in the A^+ form, that is, it is dark. On the contrary, a high-intensity signal may be a warning about bright light. Fast-changing signal intensity might mean that the light source is moving away or towards the subject. The brain processes the signal, interprets (i.e., perceives) it, and, on the basis of information stored (either instinctively or through acquired experience), decides whether there is reason for the organism to react. If it is determined that the organism must react – for instance, by fleeing – new centrifugal signals are transmitted to the muscles.

Information on the quality and quantity of light is transmitted to the brain, where…

… it is processed on the basis of information already stored…

… and a decision is reached regarding a possible reaction

So far, light has been discussed in terms of intensity – whether there is more or less of it (i.e., quantitatively) – as in the case of a colour-blind individual who perceives only shades of grey with white and black lying at the two extremes. Many animal organisms are satisfied with that; they are colour-blind and have no apparent need to perceive the quality of light, that is, colours. Others, such as humans, are capable of perceiving colours. The molecular basis of this distinction is related to the presence of different photoreceptors, each one being aroused by a different wavelength of the visible spectrum. Primates, including humans, have three types of photoreceptors, each of which is sensitive to the blue, green, or red region of the spectrum. For red light to be perceived, for example, the number of "red" photons must be greater than that of the "blue" or "green" region

Human eyes have three photoreceptors

of the spectrum. In other words, the perception of colour depends on the competitive relationships among the three photoreceptor types. If the number of photons in each of the three spectral bands is approximately the same (e.g., as in the case of direct sunlight or the light from an ordinary lamp), then all three types of photoreceptors are excited equally, none dominates, and the final perception is that of white light. Not all animal organisms have these three types of photoreceptors, and some have different types. Some insects have an extra photoreceptor sensitive in the ultraviolet region (λ at around 360 nm), thus extending the perceptible spectral region of the light environment to shorter wavelengths. Others lack receptors sensitive to the red region, meaning they do not perceive what we consider red. Generally speaking, natural selection obviously has favoured the production of photoreceptors with optical properties (i.e., the absorption range) that serve the vital interests of each organism. For example, leaf-eating insects need to distinguish fine shades of green in leaves so that, at a distance, they can discern which ones are suitable for consumption or for laying their eggs. Therefore, their photoreceptor system is particularly sensitive to the photons of the green region of the spectrum.

Not all organisms see in the same spectral region: small changes in the molecular structure or the number of photoreceptors change the visible form of the world

Because the light that comes from the only natural source that lights the earth – that is, the sun – is white, why is our environment so rich in colour? Perhaps the previous discussion has made it clear that the light finally reaching our eyes is not direct light, but light that has been reflected many times on surrounding objects. For example, the light falling on a leaf is white, but the reflected and transmitted light (i.e., that which reaches the human eye depending on the relative positions of the light source, the leaf, and the eye) is green. In other words, it contains many more photons in the green than in the other regions of the visible spectrum. Hence, it may be concluded that the interaction of

Why does the colour of objects depend on their chemical composition?

light with the leaf changed the light quality; this is because absorption is selective. Leaves contain chlorophylls, which absorb most photons of the red and blue regions. Consequently, these photons are removed from the total and the light reflected or transmitted is relatively depleted of red and blue and relatively abundant in green. The same is true of any object that contains substances with a selective capacity to absorb specific regions of the spectrum. In other words, the colour of each object depends on its chemical composition.

What do Plants 'See'?

If a layperson were asked whether plants see, his or her immediate response would be an emphatic no. Plants do not have sensory organs or nerves to transmit signals or a central coordinating organ (brain) to process and modulate the signal received. The same answer might be given about other environmental stimuli, whether chemical or physical, based on the same anthropocentric approach. However, if the question were modified slightly to ask whether organisms lacking apparent sensory organs and a nervous system could, in some way, perceive environmental stimuli, quantify them, and continuously record their changes, the answer might not be so categorically negative. However simple an organism might be (or appear to be), it could not survive without recording in detail the aspects of its environment vital to its growth, survival, and reproduction. The most illustrative case for teaching purposes, as well as the best-studied one among plants, has to do with perception of the light environment (i.e., plant vision) and the calculation of time based on this perception. In the following analysis, we maintain the basic lines adopted in the preliminary discussion of light and colour

Light perception without sensory organs and a central coordinating organ?

Survival, for any organism, would be impossible without perception of the environment

A basic sequence of reactions: collection and processing of information, decision to modify behaviour

perception: perception presupposes the absorption of light by one or more types of photoreceptor. Stimulation of the photoreceptor sets in motion a sequence of secondary biochemical changes, the final outcome of which is a developmental response by the organism or a change in its behavior. The response or behavioural change must have some adaptive value to justify the cost of constructing and maintaining the perceptive mechanism. In other words, it should be beneficial to the organism.

How do plant seeds measure light…

The seeds of many plants can germinate in absolute darkness, provided they get wet. In other plants, however, hydration is not enough; some light also is required. All it takes is a short sparkle of white light for biochemical processes to be set into motion and lead to germination, that is, the appearance and development of the young root and shoot. In other cases, the duration of stimulating light must be longer (or the number of sparkles greater). As in every other case in which light affects an organism, in the case of seeds, the first question raised concerns the nature of the photoreceptor, because for the light to have an effect it obviously must be absorbed. The quest for a photoreceptor starts with an experiment that is easy to conceive but difficult to perform. Seeds are exposed to monochromatic light, that is, light from which all wavelengths (i.e., colours) except one have been removed. Optical devices called monochromators separate white light into its constituent colours; airborne raindrops might function as such a device: when sunlight falls on them at a specific angle relative to the observer, a rainbow appears in the distance. The best-known monochromator in history is Newton's glass prism, which this eccentric scientist placed along the path of a sunbeam in a dark room to observe the separation of white light into the colours comprising it and their projection on the opposite wall. Modern monochromators (which are not prisms) are capable of analysing nanometer fractions. A seed exposed to light of a different quality (i.e., colour) may or may not germinate, depending on the capacity of its photoreceptor to absorb the specific light (i.e., the photons of the specific wavelength). In other words, what is created is an *action spectrum*, a two-dimensional graph (*X* and *Y* axes) showing the degree of the seed's response to the wavelength of incident radiation (i.e., the radiation hitting its surface). Obviously, this spectrum's peaks and valleys must coincide with those of the photoreceptor's absorption spectrum. If the seeds' germination rate is highest when they are exposed to red light, then the photoreceptor must most effectively absorb red light of the same wavelength. Consequently, experimentation continues to the next phase: now candidate seed biomolecules are isolated to detect their absorption spectrum and compare it with the germination action spectrum.

... and discern colours?

Around 1960, after a decade of hard work, the responsible photoreceptor was identified and named *phytochrome*. The phytochrome is a complex organic molecule made up of an open chain of four pyrrole rings (a substance roughly similar to heme and chlorophyll) bonded to a protein. The phytochrome is stimulated to the maximum degree with red light of around 660 nm. When this happens, its structure changes slightly and reversibly. The slightly modified phytochrome does not absorb light at 660 nm but rather at 730 nm, that is, in the infrared region (not "visible" to the human eye). The presence of the modified phytochrome form activates the molecular and biochemical mechanisms necessary for seed germination, meaning the modified phytochrome – having absorbed red light – is the active form. When the active form absorbs photons of the infrared region of the spectrum ($\lambda = 730$ nm), it returns to its inactive form, which keeps the seed dormant. Furthermore, the active form becomes inactive in the dark. Therefore, the phytochrome system operates like a valve: it opens (allows germination) or closes (favours

The most common photoreceptor in plants is roughly similar to heme and chlorophyll but has a totally different function

In essence, the photoreceptor called phytochrome counts the number of photons in the red and near-infrared regions...

... operating like a molecular valve that allows germination or extends dormancy

dormancy), depending on the number of photons in the red (around 660 nm) or the near-infrared (around 730 nm) region of the spectrum. When "red" photons prevail in the seed's environment, the seed germinates. When "infrared" photons prevail, the seed remains dormant. It also remains dormant in the absence of both types of photons, that is, in the dark. In other words, the phytochrome is a detector (or sensor) of the ratio of photons from these two regions of the spectrum. What adaptive value might be attributed to seeds being equipped with a photon detector in two regions of the spectrum, and why is this necessary? How does "knowing" the light environment in this spectral region, around 650 to 750 nm, serve the plant? Why do seeds choose to germinate when the sensor detects a prevalence of red over infrared photons?

Why do small seeds need to measure light, whereas bigger ones germinate in the dark as well?

Several decades have passed since the observation that large seeds (e.g., those of beans) may germinate in the dark, whereas tiny seeds (e.g., those of lettuce) need light. Therefore, the photon sensor apparently is necessary for small seeds, whereas it might be useless in bigger ones. How is the small seed size related to the need for light detection? To relate the two, a short digression is necessary to briefly describe the structure and some basic physiologic functions of seeds immediately before and after germination. Before a mature seed is separated from the mother plant, it is made up of a relatively hard external waterproof shell constructed of dead tissues. Within the shell are two discernible structures: the *embryo*, with the initial traces of a tiny root and a tiny shoot, and the so-called *endosperm*. The latter is the most voluminous part of the seed and contains large cells full of stored compounds (e.g., starch, fats, and proteins in various ratios, depending on the plant). When germination starts, the initial root and shoot growth needs energy and structural blocks, which the young seedling cannot absorb from the environment because its photosynthetic apparatus has not

What does a seed consist of?

Seed development and the role of stored compounds

yet developed; therefore, these necessities are provided by the endosperm. Stored substances are broken down, and the resulting micromolecules (e.g., sugars, fatty acids, and amino acids, depending on the nature of the stored compounds) are either transformed into the appropriate structural blocks to contribute towards developing new cells, tissues, and organs or are broken down further to release the energy they contain. The bigger the endosperm (or the more voluminous the seed), the more stored substances it contains and, therefore, the longer the period the young seedling is adequately supplied before it acquires photosynthetic independence and can synthesise its own food through photosynthesis. When a seed lands on the ground, it might remain on the surface or become buried under the soil, at a greater or lesser depth. If a small seed found itself very deep in the soil and attempted to germinate, the food supply of the endosperm likely would not suffice until the first leaf emerged aboveground. Such a seed would be doomed to fail and would die from lack of food. On the contrary, if the seed were quite large, its survival would be much more likely because it would have more time for germination because of the relative independence secured by the rich endosperm. Therefore, a sensor detecting light and providing information about the position of the seed increases the likelihood of its survival. The operation of such a sensor is undertaken by the phytochrome system.

Why is a small seed doomed if it germinates far below the soil surface?

Phytochrome, therefore, provides information about the position of the seed…

At this point, another question may be raised. If the only requirement is to detect the position of the seed (on or below the ground surface), why would a photoreceptor active between the visible and near-infrared spectral regions be selected instead of one active in some other region? Furthermore, why was the photoreceptor selected to detect the ratio of photons from two wavelength regions rather than the absolute incident radiation of light in one wavelength region?

… that is, about whether it is under the surface of the ground or under other plants

Plants that love light and others that prefer shade manage to cover all available space

Photosynthetic demand for light differs from plant to plant. Some plants grow comfortably and thrive in the deep shade of other plants; these are said to tolerate or seek shade. Other plants need ample light and do not grow well in the shade. The seeds of a *heliophile*, that is, a sun-loving plant, might fail to establish themselves, even if they are on the ground surface, when they are shaded by other plants, because of the relative absence of intense light. In these cases, a sensor is necessary to provide information as to whether there are other plants above the seed.

Light that passes through the foliage is light unabsorbed. Photosynthetic pigments (i.e., chlorophylls and carotenoids) in the leaves absorb intensely in the violet/blue (400–500 nm) and orange/red (600–700 nm) regions of the visible spectrum. Therefore, the light passing through is rich in green and lacking in other colours of the visible region, including red. Near infrared (700–800 nm), however, is not absorbed and passes through the foliage. Therefore, the light that penetrates is rich in this wavelength, which of course is invisible to the human eye.

If the human eye could perceive the actual distribution of wavelengths in the understorey of a forest, it would see a different world. This is the world the seed must "see" to know whether it is positioned under other plants. As already mentioned, however, seeds can indeed detect photons of this region through the phytochrome system. Remember that when infrared photons prevail over red ones, the seed does not germinate. The seed will germinate when its photoreceptor detects red light photons, that is, when overlying leaves fall or become sparse, allowing full sunlight to pass through. Indeed, the seeds of many plants that require strong light intensity to photosynthesise adequately remain dormant while the leaves of the plants above them are in place. When these leaves fall, the phytochrome is activated and the seeds below germinate and grow quickly. In essence, the seeds exploit the window that is open for a few months a year and race to grow before the plants above shoot new leaves and close off the light. Alternatively, the race concerns such plants' efforts to conclude their life cycle within the time the light window is open. They are necessarily plants with a short life span.

Exploitation of the winter light window in the forests

In contrast, the seeds of plants that tolerate or seek shade have no reason to wait. Their phytochrome system is not as strict and allows germination even when the red/infrared photon ratio is low.

The next question might well concern the mechanisms through which initial light absorption by the photoreceptor is translated into the final result. What takes place between light absorption and behavioural change? What is the force and what is the mechanism through which the active phytochrome induces germination? How is the signal transmitted? Here, things are more obscure. There are several hypotheses, as well as corresponding indications to support them, but no clear picture of the sequence of events from the detection of light to the final result, that is, germination. Therefore, to help readers understand what is meant by signal transmission, the prevailing hypothesis will be addressed. According to this hypothesis, detection of the appropriate red/infrared photon ratio by the photoreceptor activates a calcium (Ca^{2+}) pump, which allows this element to be transported from outside the cell (cell wall) to the cytoplasm (i.e., the area of the cell that appears to have no form under an electronic microscope and contains cell organelles such as the nucleus, mitochondria, plastids, and vacuoles). The increase of Ca^{2+} within the cytoplasm activates new pumps in the membrane separating the vacuole from the cytoplasm, further increasing Ca^{2+}. Calcium activates proteins that regulate gene expression; in this way, genes related to and necessary for germination are "unlocked" and produce the needed products. These products may include

The light detector measures the red-to-infrared photon ratio...

... and "opens" calcium pumps in cellular membranes...

enzymes that act as catalysts in the biosynthetic pathway of the hormone responsible for releasing seeds from dormancy. The sequence of events is as follows:

. . . while calcium, in turn, activates proteins. . .

. . . that awaken genes responsible for the onset of the developmental programme of seed germination

- There is a change in the light environment indicating that external conditions are appropriate for the successful establishment of the young seedling (e.g., the seed is in a position with sufficient light).
- The photoreceptor (phytochrome) detects the quality of light and, like a switch, opens the valves for the transportation of calcium into the cell interior.
- The calcium reacts with special proteins and enables them to enter the nucleus and reach the genes containing the information titled "developmental programme for seed germination."

This information is transmitted to the sites where antidotes to dormancy and compounds promoting germination are synthesised.

The role of the phytochrome is not limited to detecting the seed environment. . .

The role of the photoreceptor is not limited to locating the seed's position. Like a sleepless eye, it continues to detect the position of the growing plant and to send signals for a faster or slower growth rate as well as for metabolic rate changes in either direction. In essence, the photoreceptor is the intermediary that guarantees that internal growth processes are coordinated with the changing external environment. Therefore, a seed that germinated, say, 1 to 2 cm below the ground surface must unfold its first leaf immediately so that it soon emerges from the surface. Upon emergence, however, the light quality is totally different and interpreted as information relating to "finally, I am out." There no longer is concern about longitudinal growth. The plant can slow down a bit because now it is more important for it to develop its photosynthesising capacity quickly, to become independent before the endosperm's supplies are depleted. What is necessary at this point, then, is not vertical but horizontal growth; in other words, the leaf surface must expand so it captures more light.

Internally, metabolism needs to be modified so that the chloroplasts' photosynthetic membrane constituents and corresponding enzymic equipment start to be synthesised to assimilate CO_2. The ever-alert phytochrome, however, also provides warning about the likelihood of other, taller plants nearby, which might hide the sun. If a great amount of light is necessary for the specific plant, the signal is translated as a mandate for faster upward growth so that it may outgrow its rivals. In this case, it is not enough just to emerge from the ground surface; the plant must grow above the foliage of its rivals in a never-ending battle for exploitation of resources. Of course, there also are undemanding plants that settle for less light, at the cost of a smaller size, to avoid brutal competition. In these plants, the same signal detected (i.e., the same red/infrared photon ratio) is interpreted as a signal for moderate upward growth. Undergrowth plants make do with the leftover light and lead a quieter life, free from the need to continuously modify their height to keep up with their more aggressive neighbours.

...it also provides information on the light (i.e., energy) conditions throughout the plant's life

One's neighbours have the potential to be both competitors and protectors. A plant and its neighbours compete for space, food, light, water, and minerals in the soil. However, the neighbours also are protectors because their presence makes an individual plant less apparent. A predator (e.g., a herbivorous animal) may have its fill by nibbling a little of each inhabitant; it might reach satiation by eating small bites of each plant without exterminating any of them. Chapter 7 includes an examination of the methods plants use to encourage or discourage other plant species from establishing themselves nearby; it shows how plants select their neighbours and build "social relationships." For now, the question is whether plants, which are aware of other plants above them, are equally aware of other plants around them. A neighbour of the same height is a potential rival for light if its growth rate is faster.

How do plants become aware of the presence of neighbours (i.e., competitors)...

... not only when these neighbours are above them, but also around or below them?

Therefore, it has been discovered that plants, through the same photoreceptor – that is, the phytochrome – detect the presence of likely neighbours and modify their height accordingly. The light that stimulates them, however, is not the light passing through the leaf but the light reflected from it, the light unabsorbed by chlorophylls – in other words, green light (visible to the human eye) and infrared ("visible" to the phytochrome) radiation. Consequently, the light reflected from a leaf contains far fewer "red" photons (because they have been absorbed by the chlorophylls) and many more infrared ones. As for undergrowth plants, which love intense light, the light quality (i.e., "colour") also is detected by the phytochrome and triggers reactions that tend to increase height. Neighbouring plants, of course, respond the same way. In nonstop bidding for the prize of light, everybody keeps growing taller until something else limits further upward growth. This obstacle may be difficulty in raising water, the wind, or an inability to support greater weight. In other words, at a certain point further competition favours nobody, because what is gained in light as a result of greater height may be outweighed by a limited water supply or a greater need to invest in support mechanisms. At this point, everyone seeks a compromise, which, in the case of a forest, is expressed as a mean tree height with minor deviations, in which no individual sticks its head up. The absolute height value depends on the potential of the prevalent tree species to exploit the specific environment.

Potentially competitive neighbours, growth rate, and a race for the prize of light…

…however, it all comes down to a compromise in the end

In light of the preceding description, the question of whether plants "see" may be raised again. According to dictionaries, vision is the perception of light through the eyes. What does not seem fitting in the case of plants, however, is the term *eyes*. Indeed, throughout the previous discussion, there was no mention of the photoreceptor called the phytochrome being located in a particular

"Vision" in plants is diffuse in almost every plant cell

organ with a specific structure that might be characterized as an "eye." On the contrary, the phytochrome appears to be present in every living cell of a plant. On the other hand, based on what has been described, plants cannot be considered "blind," that is, lacking in vision. Because all plants "see," they meet all the prerequisite conditions stipulated by the definition of *vision* except for the presence of eyes. This diffuse "vision" may be related to the absence of a central coordinating system that controls behaviour, combined with the relevant independence of a plant's constituent parts. Each branch must find its own way, each detecting in its environment the photon ratio indicating the appropriate position. A single organ perceiving light, however centrally it may be located, could not successfully assess the light quality at every nook and corner of an architecturally complex organism that keeps growing. Furthermore, a central information processing unit would be necessary to classify and transmit signals.

Equally diffuse is the processing of messages in the absence of a central coordinating organ

Therefore, through the phytochrome system, every plant cell apparently is capable of perceiving light quality within a critical spectral region and of transforming the stimuli into a sequence of signals, allowing it, finally, to respond appropriately to the initial light information. Although this description may not meet the definition of *vision* from an etymologic point of view – and plants do not have eyes or a glaring, vague, innocent, or guilty look – plants are capable of perceiving light quality, transmitting signals, and modifying their behavior accordingly. In other words, they perceive a light stimulus and therefore have the corresponding sense.

In what other ways do plants use their unique "vision"?

How do Plants Measure Time?

Why do plants have to measure time?

Like all organisms, plants need to measure time, particularly so because they are immotile organisms, that is, they are at the mercy of environmental conditions; therefore, they gain in adaptability if they can anticipate imminent natural seasonal changes. During the adverse season of the year, which in cold northern regions is winter, perennials in these latitudes must be prepared to face rough conditions; for example, they must shield the cells of their leaves with appropriate mechanisms to prevent ice crystals from forming in their interior. If a plant cannot achieve this, it may shed its leaves before winter comes. Leaf ageing and shedding is a programmed, controlled, and active process, the function of which is to transport the useful nutrients contained in the leaves to the rest of the plant body, which is ready to go into dormancy. Therefore, this process must occur well before low winter temperatures prohibit the metabolic events necessary for such transportation. For this reason, leaf ageing starts much earlier and obeys environmental signals indicating the advent of autumn. The plant must perceive these signals through appropriate sensors and translate them into a specific developmental response: leaf ageing and shedding. It may seem strange to consider leaf ageing a developmental response; however, as explained earlier, *ageing* and *death* in plants have a totally different meaning than they do in animals. Every deciduous species has its own 2- to 3-week period, generally the same time every year, during which it sheds its leaves. How does it perceive time?

The example of coordinated leaf shedding

Keeping the wolf from the door, or why leaves fall before the adverse season arrives

A second developmental process, much more precise with regard to time, is flowering. Consider a meadow that is green one day and turns red overnight, when all the poppies decide to bloom in sync. Indeed, many plants bloom not only during a specific season but (within some acceptable margin of variation) on a specific date, with almost clockwork punctuality. The necessity and adaptive value of such coordinated seasonal behavior are discussed further later. In the meantime, we will explore how plants perceive the passage of time.

Imagine a person has fallen into a long, deep sleep over an undetermined length of time and suddenly wakes up. What would this person need to determine today's date? If he or she were told not to look at a calendar or newspaper, or to turn on the TV, or to ask the audience for help, what could he or she do? Of course, he or she could look around to identify environmental parameters to indicate the season. If it is warm or cold, it might reasonably be assumed it is summer or winter, respectively. However, what if it is a warm or cold spring day? In other words, temperature may be misleading. In contrast, the duration of daytime (or nighttime) would be a more accurate criterion, because every 24-hour diurnal cycle within a year has a fixed day and night duration, respectively. For example, counting 10 daytime and 14 nighttime hours (based on the times of sunrise and sunset) would provide a more useful piece of information. However, this approach also might be confusing because that specific period of daylight (or nighttime) occurs twice a year, once in the spring and once in the autumn. Therefore, the duration of day (or night) would have to be measured on at least two consecutive days to determine whether the calendar is moving towards winter or summer. This information would be helpful in preparing for the coming season (e.g., buying the right clothes). This is exactly what most plants do: they measure the length of daytime (and nighttime) and perceive the time passing and the progress of seasons. Technically, it is said that plants measure the *photoperiod*. For every plant, a critical daytime duration has been selected (with the corresponding critical nighttime duration); if this is

How can we calculate the date without a clock or calendar?

Plants measure the duration of day and night daily

This is how they perceive the progress of the seasons and "decide" when to flower

exceeded, functions are activated that result in a developmental process, such as flowering. The critical duration for each plant has been determined and entered into its internal developmental programme so that it is the most appropriate for the potential of the plant in question (with its given reproductive biology) and for the specific environment. In many plants, the photoperiodic stimulus is insurmountable and absolutely set; in other words, they do not flower if they do not receive it. This is why many plants from other parts of the world do not flower in Greece. For example, a plant brought to Greece from Norway, where the plant bloomed in June, when daylight there exceeds 20 days, would never flower in Greece because it would never encounter such long days. The only way to make the plant flower would be to extend daylight with artificial lighting.

Why do some plants from other parts of the world not flower in Greece?

So, what type of clock do plants use to measure the duration of night and day? The phytochrome also seems to be involved here, as in the case of seeds germinating in response to light. Remember that phytochrome is a photoreceptor that exists in two forms, active and inactive, and their ratio depends on the quality of light the plant receives, specifically the ratio of red to near-infrared photons. During the day, regardless of weather or other conditions, photons of both spectral regions stimulate the phytochrome, restoring a ratio between its two forms that contains a sufficient quantity of active phytochrome. In darkness, however, the active form quickly shifts to the inactive side through a biochemical process. Therefore, with regard to the phytochrome, a day in the life of a plant includes two periods: the light one, during which part of the photoreceptor is active, and the dark one, during which the entire photoreceptor is inactive. In other words, the basis of the plant clock is the phytochrome. Yet, how is this relative way of measuring time translated into a signal? Imagine the phytochrome tipping over an hourglass. The hourglass is

The phytochrome participates in measuring time

A molecular hourglass

tipped over at sunrise and sunset, depending on the prevailing phytochrome form. The sand accumulated during the day is removed at night; the longer the day, the more sand is accumulated. Imagine also that when the quantity of sand exceeds a critical value, this triggers a process that inevitably and irreversibly leads to flowering, as though the support gives way under the weight of the sand.

In more technical terminology, the active form of the phytochrome induces a biochemical pathway that produces a compound that accumulates over time. When the concentration of this compound exceeds a critical value, it elicits the response of the flowering genes. These genes contain information about growth regulators (hormones) specific to this process that are produced in the leaves. From the leaves, the hormones are transported to the flower buds, bidding them to develop.

In other words, the photoperiod winds the biological clock; therefore, it is very accurate – all individuals of a species within a location receive the initial flowering signal on about the same day. However, the date the last flowers appear on the plants of a certain location varies depending on the specific spot a plant is located. For example, poppies growing high on a mountain slope, although belonging to the same species as those at the base, flower a few days or weeks later, depending on their altitude. However, this does not mean they received the photoperiodic stimulus later. They received it the same time as the poppies at the lower altitude, because they all are at the same latitude. Why the delay, then?

Why is the flowering season delayed at higher altitudes?

The duration of day and night does not depend on temperature. The signal is to be received on the same day (or days), regardless of whether the day is bright or rainy, hot or freezing. The sensitivity of the sensor does not depend on the weather. However, the subsequent process is biochemical, and biochemical processes are

Clock sensitivity is not dependent on temperature

highly dependent on temperature. Therefore, if the prevailing temperature is low after the photoperiodic stimulus is received, the final outcome of flowering is delayed. If the temperature is high, flowering occurs sooner. It is as though the triggering always takes place at the same time point, but the bullet covers the distance to the target at different time spans. In other words, the time the process begins is absolute, but the time it ends (i.e., flower deployment) is relative, shortened, or lengthened depending on the temperature, which determines the speed of biochemical reactions, that is, the time it takes to achieve the final goal.

The time the process begins (reception of photoperiodic stimulus) is constant, but the time it ends (appearance of flowers) depends on the temperature

Does this shortening or lengthening of time have any adaptive value? The plants would be in trouble if it did not! Flowering must occur concurrently with peak pollinator activity, as was shown in Chapter 5. From a reproductive point of view, it would be a pitiful failure if the plant flowered too early or too late relative to the peak activity time of the intermediary (insect, bird, or mammal) that brings into contact the plant's reproductively critical male (pollen) and female (ova) cells. Because the growth and activity of intermediaries depend on temperature as much as flower development, temperature undertakes the coordinating role and the concurrence of the two activities.

This is how flowering is coordinated with the activity of pollinators

What else do Plants Perceive?

Focusing on light alone, one may now ask whether plants are equipped with photoreceptors other than the phytochrome. It is important to keep in mind that this discussion concerns the use of light as information, not as energy. The latter entails chlorophylls and the photosynthetic function.

How do plants turn to light?

Two cases have been studied thoroughly. The first one is common experience: anyone who has ever planted lentil seeds on a window sill has observed that young shoots turn towards the light. The reason is obvious: turning to the energy source ensures maximum exploitation. When light comes from a specific direction, the plant reacts by decelerating the growth of cells on the lit aspect of its stem so that the stem bends to the light, forming a lit concave side and a shaded convex one. The light receiver involved, called *phototropin*, "sees" blue light, that is, it absorbs in the blue region of the visible spectrum. Phototropin is a protein with incorporated flavin, the compound responsible for the absorption of "blue" photons. As in the case of the phytochrome, the sequence of signals following light absorption entails calcium, which in turn activates the protein complex–receptor of a plant hormone called *auxin*. This way, the concentration of auxin is altered differently in the cells of the two sides of the plant stem, which react by regulating their final size so they are bigger on the shaded than on the lit side.

Blue light receptors, plant hormones, and asymmetric growth

The second case is not visible to the naked eye but may be observed using specific instruments. Plants photosynthesise only during the day, when there is light. The gates for the entry of CO_2, which will be reduced to sugars with the help of photosynthesis, are the stomata, small openings on the leaf surface flanked by two special cells called guard cells. These cells swell and shrink to regulate the size of the opening. Inevitably, however, it is not only a question of CO_2 entering through the stomata, but also of water vapours exiting. At nighttime, when there is no photosynthesis, the stomata close to prevent dehydration, whereas during the day they are open to allow photosynthesis. The rationale for the function of these valves, which are critical for the life of a plant, may be expressed in financial terms. Obviously, at night they have to remain closed; however, even during the day, the size of the opening must be regulated so the plant may derive the maximum benefit from photosynthesis without risking

To regulate their final opening, microscopic plant stomata detect and process three environmental signals…

dehydration. In other words, it is no coincidence that guard cells regulate the valve opening, continuously recording three critical environmental factors: the relative humidity of the atmosphere, the concentration of CO_2 inside the leaf, and the light intensity. The size of the opening depends on processing the values of these three parameters.

... the relative humidity of the atmosphere...

When the relative humidity of the atmosphere is low (i.e., the air is relatively dry and, therefore, the tendency for evaporation is greater), stomata tend to close so precious water is not lost. When CO_2 is reduced in the leaf interior, the stomata tend to open. A low level of CO_2 in intercellular spaces means that intense and effective photosynthesis is turning it into sugars. Therefore, it is reasonable for stomata to open wider so CO_2 may be channeled from the atmosphere to the leaf interior to replace what has been assimilated. These two factors are responsible for the fine adjustment of stomata opening. However, the primer for the drastic change from tightly closed stomata at night to (more or less) open stomata during the day is light. The photoreceptor responsible is a carotenoid (zeaxanthin) that is contained in the chloroplasts of the guard cells and absorbs blue light. The activation of zeaxanthin triggers reactions that result in a deformation of the guard cells, which opens the stomata. A full description of these reactions is beyond the scope of this book. However, it is important to keep in mind that guard cells contain sensors to measure three characteristic parameters – light, humidity, and CO_2 – and are equipped with mechanisms to process the values obtained. If the values are favourable (plenty of light, a low CO_2 level, and a humid atmosphere), maximum stomata opening is achieved. If some values are favourable and others are not, the final processing results in a compromise of contradicting effects to obtain the best opening size to respond to the set of environmental conditions at any given time. The aim is

... the concentration of CO_2 inside the leaf...

... the presence of light

Processing the values collected by the three sensors determines the final size of the opening to ensure optimal photosynthesis at minimal water loss under the given circumstances at any point

to achieve the highest level of photosynthesis possible with the least amount of water loss. Of course, these activities occur to the extent that survival is not at stake. To ensure the plant is not at risk for dehydration, the guard cells also receive signals from the root informing them about the humidity of the soil. This is particularly important because atmospheric conditions might be suitable for stomata opening (e.g., sufficient light, humidity), but the soil might be dry. In this case, the root produces a plant hormone (known as abscisic acid), which is transported through the xylem vessels to the leaves and, ultimately, to the guard cells. This hormone inactivates the stomatal opening mechanism, nullifying all previous effects of light, humidity, and carbon dioxide. It is an emergency signal indicating the inability of the root to supply water to the leaves. When faced with this life-threatening emergency, interrupting photosynthesis is less significant.

A fourth hormonal signal from the root countermands the function of the sensors and closes stomata tightly, when there is no water in the soil

The next reasonable question that might be raised is whether plants are capable of perceiving environmental stimuli other than light. One way to approach this issue is to examine which environmental stimuli might be useful and whether their perception would provide plants with some adaptive or comparative advantage. In this way, some stimuli would be *a priori* excluded, whereas others would be candidates for further research. For example, animals use hearing to perceive imminent danger or to locate their potential mate from a distance. In both cases, they respond by moving towards or away from their target. Plants cannot react by fleeing in the presence of an enemy, nor can they approach their mate, however aroused they might be. As seen in Chapters 5 and 7, various clever ways of self-defence or "mating" contact have emerged, ways that befit an essentially nonmotile organism. Perceiving sound waves (i.e., fine fluctuations in air pressure) probably does not provide an advantage for plants. It is often said that plants respond and grow better when they "listen"

Plants do not perceive sound waves (in other words, they do not "hear"), nor do they like music to specific types of music. Despite the publicity such views enjoy from time to time in the popular press (and the ease with which the average person believes them), plants have no musical preferences. Humans, however, not only have preferences but also maintain illusions. A careless Western experimentalist, a fan of classical music prone to sensationalism, would probably claim that plants grow better when exposed to the soft sound of a Bach sonata. A similar person from the East might claim the same about an explosive and sensuous belly dance tune. An impartial biologist would claim that, at least to date, there is no convincing evidence that plants perceive sounds. Similarly, plants have no reason to perceive the chemical composition of their food through something similar to humans' sense of taste. Plant food is extremely simple, lacking variety or hidden poisons. All a plant needs is CO_2, water, light, and a handful of inorganic mineral salts taken directly from the environment and not from another organism, which in an effort to defend itself might excrete toxins.

However, they do perceive the presence of some volatile substances in the atmosphere, vibrations, and occasionally physical contact

Chapters 7, 9 and 10 contain references to plants' perception of volatile substances in the atmosphere (similar to the olfactory sense in animals), vibration (shock), and, curiously, direct contact (similar to the sense of touch in animals). For now, the chapter on plant perceptive mechanisms is closed by referring to the perception of gravity.

Something quite useful: the perception of the direction of gravity, or why roots are directed towards the centre of the earth

What is known from common experience is that the root always moves downwards. Even if a young plant were turned around by, say, 90° and rendered horizontal, after a few hours the root would have turned in the direction of gravity. This is a natural tendency, because the root is assigned the task of absorbing water and inorganic minerals from the soil. Of course, it must be stressed that the final angle of the root in relation to the direction of gravity is a result of the combination of the response to gravity and other soil stimuli, such as the probably uneven distribution of resources in the soil. With regard to gravity, though, a distinction should be made among three successive stages: perception of the external stimulus, generation of the internal message, and translation of this message into a developmental response: a higher growth rate in the upper than the lower side of the root so the corresponding bend may be created.

What, then, is the sensory mechanism involved? At the tip of each root is a population of cells that make up a tissue called the *root cap*. In the centre of the cap is a group of cells called *statocytes* with voluminous internal organelles called *plastids*. Plastids contain sizeable starch granules. Starch plastids, or *amyloplasts*, always arrange themselves in the lower part of the cell because their specific weight is greater than that of other cell constituents. Surgical removal of the root cap causes the root to lose its orientation, that is, it makes it turn randomly to various directions. Therefore, gravity sensors are located in the root cap. An artificial change in the direction of gravity causes an almost immediate (within 5 minutes) shift of *amyloplasts* to the new "lower" end of the cell. In other words, there are strong indications that *amyloplasts* are related to gravitropic perception. In a nutshell, the probable sequence of events is as follows: Amyloplasts, because of their greater specific weight, precipitate to the bottom of the cells. They also are connected to one another and to cellular membranes through a network of contractile protein microfibres. Their unilateral precipitation leads to tension (stretching) of the contractile proteins, which transfer this tension to the external cellular membrane and the membranes of cellular organelles. Contractile proteins share many features with the corresponding contractile proteins of animal muscles. Intense research is being conducted on how this tension generates a message, and how this message is transmitted and delivered to cells at a distance from the cap so they may enlarge to a greater or lesser extent and create a bend in

The specific weight of cellular organelles as a basis for the perception of gravity

The role of elastic protein microfibres, similar to those of muscles...

. . .and the involvement of a plant hormone

the root. It seems, however, that the phenomenon entails the synthesis and action of a plant hormone called auxin, as well as the gates transporting this hormone from cell to cell.

To conclude the topic of environmental stimuli detected by the roots, let us briefly examine the patchiness of the soil matrix. Soil is a complex mixture of solid particles of various sizes, created as a result of mother rock detritions; living microorganisms and bigger soil-living animals; plant roots; organic matter derived from dead organisms or their parts; organic secretions; and excreta. Another constituent of soil is water, which contains dissolved organic and inorganic compounds. The whole system is highly patchy, with significant local fluctuations in the concentration of resources plants need to exploit. For example, how does a root react when, in its advance, it has to penetrate the buried excreta of a cow? As disgusting as this idea may seem, the root's contact with such excreta is not at all appalling for the plant; one may observe the positive results of adding manure to soil under cultivation. Animal excreta contain large quantities of phosphate and nitrate ions. These substances are necessary for plant growth and, *mutatis mutandis*, they are a wonderful treat for plants. Therefore, the root's contact with this localised accumulation of nutrients induces the creation of hundreds of side rootlets with the obvious purpose of increasing the absorbing root surface. It has been found that meristematic cells (embryonic multipotent cells found not only in the root but throughout the plant body) are equipped with an unknown number of sensors to detect such a localised increase in nutrient accumulation. After such detection, plants synthesise a message compound, which is transported to the cell nucleus and activates dormant genes. These in turn produce secondary messages ordering cellular divisions, which ultimately result in the creation of side roots to exploit such a precious finding.

How does a root react when it encounters animal excreta?

Sensors of nutrients. . .

. . .and awakening of genes that change the root structure

However, roots do not always have a clear path to advance. What happens if the root's path is blocked by a stone? The root would not stop growing, rather it would embrace the rock. However, this motion violates the general principle of the root's downward growth. Therefore, the results of the gravitropic stimuli must be countermanded so that growth may deviate and follow the contour of the almost-horizontal stone surface. The stimulus the root perceives to counter the effect of gravity seems to be the pressure applied between the gravitropic root and the stone. The consequent mechanism of countering the gravitropic result is not known, but it appears to involve the production of an unusual plant hormone called ethylene. Ethylene, a very simple organic molecule comprising two carbon and four hydrogen atoms, is unusual in that it is volatile. In other words, it is a hormone that acts on the cells producing it during the period from its production to its fast diffusion into the atmosphere. In contrast, all other plant and animal hormones are produced at one location but are transported to and act only on target tissues far from the site of their production. After the root embraces the stone and becomes perpendicular again, the pressure obviously is removed and normal gravitropic function restored.

What happens if the root encounters a hard object?

The root perceives pressure. The tactile stimulus countermands the gravitropic one, so the root changes direction and embraces the stone.

What happens if the root does not encounter a stone but a huge rock? Here, the behaviour of the root of some trees and certain perennial woody bushes differs. These roots do not embrace the rock, they penetrate it. The root secretes acids and creates micro-detritions through which it reaches great depths. At the same time, rock detritions release inorganic elements for the root to absorb. In other words, the root, through its own actions (acid secretion), creates the room necessary for its advancement and changes its physical–chemical environment to exploit it.

However, it might also "decide" to break into it, if it "judges" that it can do so

Therefore, what has been demonstrated in the previous pages is that a series of environmental factors are detected by plants, providing information about the availability of resources, their position in relation to the plant, and their changes over time. Based on this information, internal signals (usually hormones) are produced and internal information is modified (i.e., the right genes awaken or become dormant) so that the plant's growth/development and behaviour is in tune with the environment. Hence, everything occurs at the right time and is produced in the right quantity and the right manner, befitting an immotile organism that is self-sufficient with regard to its nutritional and energy needs. However, our analysis so far has concerned the physical (abiotic) environment, almost totally ignoring the presence of other organisms as carriers of information that might be useful to the plant. The only reference to such organisms has been made in relation to detecting plant neighbours through the phytochrome system, whereupon the reaction of the detecting plant is to grow upwards so it may better compete with its neighbours for light. Therefore, a question may be raised regarding whether there are other cases in which it might be useful for plants to "detect" other organisms, such as microorganisms or animals. In theory, an animal organism is of interest to a plant if it is the plant's consumer. A satiated passerby is indifferent to the plant, but becomes dangerous once it begins attacking the plant. How can an immotile plant avoid being overconsumed? Or, to put it another way, how does planet Earth remain green? Experience has shown that food sufficiency leads to an increase in population, which in turn needs more food, and so on. Because there are ample plants on Earth, one might expect the population of herbivores to increase, demanding even more green food, which would reasonably lead to the extinction of plants and, therefore, of herbivores as well. Instead, there is a wonderful balance in nature between

Brief summary of plant perception of environmental stimuli

How do plants defend themselves against their enemies?

producers and consumers, between prey and predator. Therefore, it must be assumed that plants might not be the ideal food they appear to be at first sight. What protects plants from being overconsumed? How do they resist attack by pathogenic microorganisms? These questions lead to the world of plant defence.

Chapter 7
The Defence of a Stationary Organism

The Complex Chemistry of Plants Determines the Dietary Habits of Herbivores

Plants have a particularly rich biochemical arsenal

As shown in previous pages, plant pollination and seed dispersion often depend on mutually beneficial relationships between plants and animals, relationships based on chemical substances produced and sometimes released into the atmosphere by plants. This is a good point, therefore, to discuss the defence of plants against herbivorous animals, which, to a great extent, depends on the rich biochemical arsenal of plants and on the plants' ability to produce a wide variety of organic molecules. This richness, which chemists and biologists had suspected since the dawn of modern science, was confirmed in recent decades as the advance of technology significantly improved the analytic techniques and methods used by these scientists.

Plants outclass animals with regard to so-called secondary metabolism. Primary metabolism, which all organisms perform, concerns the basic metabolic patterns: synthesis of nucleic acids (DNA, RNA) and proteins, as well as the transformation of lipids, sugars, and organic acids into one another, or a slight modification of similar compounds supplied by food intake. The privilege of secondary metabolism is enjoyed by plants, fungi, bacteria, and certain protists. Secondary metabolism entails processing the products of primary metabolism via complex biosynthetic pathways and producing complex organic molecules that are not directly related to what might be called – somewhat off-handedly – the basic processes of cells and organisms. These molecules, initially called secondary plant products, are not directly related to energy metabolism, cellular respiration, enzyme synthesis, membrane construction, cell and organism movement, reproduction, genetic information or its transfer to offspring, and so on. The term *secondary products* reflects the secondary importance that was attributed to them.

The so-called secondary plant products are anything but secondary in significance

Sixty years ago, the prevailing view was that these compounds were useless by-products of primary metabolism. According to this view, plants, as incomplete organisms not equipped with an excretory system, have no choice but to accumulate these products within their cells and to tolerate them, because they cannot dispose of them. This view was reinforced by the fact that secondary products are stored mainly in the vacuoles – organelles that are indeed storage sites for plant cells – or are attached to the cellular wall, that is, outside the cell. In some cases, these substances are washed to the ground by rain or released into the atmosphere, which initially was believed to be the plant's attempt to get rid of waste. Of course, there were heretics at that time who could not easily accept that in the course of hundreds of millions of years of evolution, natural selection had not found ways to biologically use such products, even if they might have been useless initially. The fact that secondary metabolism has been under strong evolutionary pressure is indicated by the wide range of compounds produced. To date, more than 120,000 compounds have been identified, and new ones are added daily to the scientific literature. One plant species alone may be capable of synthesising several hundred of these substances. What is the purpose of the emergence of such complex biochemical pathways, such a multitude of products, such a large number of enzyme-catalysed reactions? Why should such a complex and expensive chemical factory be created just to manufacture useless products? This was a difficult idea for post-Darwinian biologists to accept. It should also be difficult for those who have read the book up to this point to accept that plants are incomplete organisms. This is how the view that secondary products must have an important role gradually started to gain ground, and the clarification of many such roles has contributed towards their nomenclatorial

The discovery of the significant roles played by secondary plant products has led to their linguistic promotion: today they are called natural plant products

These substances are the words of a chemical language for the communication of organisms, through which a harmonious ecologic balance is restored

promotion: from *secondary* they were renamed *natural plant products*. Today, it is accepted that these compounds are biochemical intermediaries in the communication and relationships of plants with other organisms and with one another. These relationships, mainly competitive and rarely altruistic, include nutritional relationships between plants and herbivorous animals. As will be demonstrated, natural plant products determine to a great extent the balance and harmony of natural populations and their relationships to one another.

Natural products have played a decisive role not only in the history of natural populations, but in human history as well. Examples are as follows:

Natural products have played a major role in human history and civilisation

- The curare used on hunters' arrows to improve the efficiency of their efforts
- The cyanide taken in gradually increasing dosages that protected Mithridates, King of Pontus, from being poisoned by it
- The hemlock Socrates drank in his tragic effort to teach the importance of observing state laws, even if they might be unfair
- The hemlock drunk voluntarily by the elderly in the squares of isolated island towns in ancient Greece, when famine threatened their communities
- Medicines and poisons
- Tranquillizers and stimulants
- Hallucinatory and aromatic substances
- Pigments and preservatives
- Spices and flavourings

The latter five were taken exclusively from plants for thousands of years, until the advent of modern chemistry. Even today, 25% of medicines in the Western world contain at least one ingredient of plant origin; this percentage increases significantly if traditional, nonindustrialised medicine is taken into account.

There is great concern over reduced biodiversity and the extinction of plant species not only for purely ecologic reasons, but also because of the utilitarian fear of losing sources of useful substances not yet discovered. Of course, few would seriously claim that plants make aromatic compounds for human enjoyment, morphine for pain relief, cocaine for its stimulatory effects, or caffeine to wake us up in the morning. These substances must be of some value, they must have a role to play for the plants themselves.

Plants, however, produce natural products for their own reasons rather than to serve humans

As already mentioned, plants, being autotrophic organisms, use exceptionally simple inorganic compounds available in their environment (water and mineral salts from the soil, CO_2 from the atmosphere) and sunlight, as an energy source, to synthesise the complex organic substances necessary for their growth and development. These properties place plants at the so-called base of the food chain, which supports the growth and maintenance of all other organisms. In theory, the availability of abundant plant biomass for food should lead to a herbivorous population explosion in the food pyramid, resulting in overconsumption of plants and a corresponding reduction of vegetation to the point of extinction. Instead, however, a characteristic balance exists in nature that indicates, among other things, that plants might not be an ideal food for herbivores. How does nature prevent overgrazing? How do plants defend themselves? Why does the earth remain green?

Why does the earth remain green?

Plants cannot defend themselves by fleeing; they have to develop defence mechanisms against herbivores where they stand. What would a host do to discourage a hanger-on from staying for dinner if the host's biologic power did not allow him or her to throw the "guest" out or if the host's tactful manners had no effect, which usually is the case with the hungry and the deprived? The host might pretend to be a terrible cook and slightly burn the dinner. If the guest persists, the host must use more effective measures. Who would return to a restaurant after breaking a tooth on a small stone in the lentil soup, or not sleeping a wink trying to digest a raw steak, or suffering a hangover because of cheap wine, or contracting diarrhea, when the initial purpose of the meal had been to enchant one's partner? Even if one had to go back to that restaurant because there was no other place to eat in the vicinity, one would make sure not to eat his or her fill. *Mutatis mutandis*, these are the types of tricks plants use; they are tricks that make plant tissues difficult to chew, tough, indigestible, and sometimes toxic.

Insects are bulimic

Insects eat a tremendous amount. From the moment they hatch until their metamorphosis (within 5–6 days), caterpillars become a few thousand times heavier. Every day, they consume a quantity of leaves equal to six times their body weight. This is similar to a 10-year-old child eating 180 kg of food daily or a regular portion every 2 to 3 minutes. Aphids (plant lice, insects whose mouth apparatus is shaped like a short, hard straw to suck juices from leaf vessels) consume juice equal to 100 times their body weight daily. In the presence of such voracious eaters, plants must develop effective defence mechanisms against overgrazing.

Mechanical defence: small stones in lentils, a raw steak, and fish bones

Leaf toughness is a matter of mechanical properties. Thorns are the most obvious expression of mechanical defence. Less apparent but equally effective is the presence of lignified, flaking cellular walls, as well as a thick cuticle (the waxy layer over epidermal cells that prevents excessive water loss). The chemical substances participating in the respective structures (cutin, cellulose, pectin) are nontoxic. However, their nutritional value is low, because they consist only of carbon, hydrogen, and oxygen. Furthermore, they are indigestible for most herbivores. At this point, it should be stressed that the main dietary goal of a herbivore is to obtain nitrogen, vitamins, and mineral salts. The last two are abundant in plant tissues, but nitrogen is not. Plant tissues contain 1% to 3% nitrogen, whereas animal tissues contain more than 10%. Therefore, for a herbivore to take in the nitrogen it needs, it must consume huge quantities of cellulose, which it usually cannot digest. Faced with the prospect of indigestion, it will eat less; however, this is not the most important aspect. After all, the herbivore's alimentary canal can absorb what is useful and excrete what is not. What is important is that to destroy the hard external "casing" so the animal can reach the valuable plant cell interior, more energy and time are consumed. It is one thing to chew on milk-fed lamb but quite another to chew on mutton from an old ewe that has wandered miles over mountain slopes. In other words, not only is cellulose-rich food low in nutritional value, but this value is diminished further by the energy consumed eating it. Besides the energy required to chew hard food, another necessity is time, during which a herbivore is exposed and vulnerable to predators. When one is being chased, there is no time to linger over menu choices or to ask the predator to wait until the lobster is shelled.

Animals are interested mainly in nitrogen

Eating dangerously

A leaf surface also provides mechanical defence through its relief, which finds its richest form in hairy leaves. One should suspect that a leaf is hairy when it is highly reflective and grey-green in colour, such as the undersurface of an olive or poplar tree leaf. Hairs usually are made up of dead cells of little nutritional value and protect the plant in many ways. They keep moisture on the leaf surface and limit water loss. They function as a sunscreen by reflecting visible wavelengths of the spectrum and absorbing ultraviolet ones. Finally, hairs impede small herbivorous insects, which have to avoid (or eat) the hairs before they can access the precious leaf interior. *Mutatis mutandis*, it is like trying to pick wild greens through thick, thorny bush foliage while being chased by an enemy.

Why do leaves have hairs?

Other plants ensure mechanical defence through the synthesis and maintenance of mineral crystals, usually calcium oxalate or carbonate, by special cells called *idioblasts*. These crystals are so hard they can destroy the masticatory organs of an unsuspecting insect – they are like small stones in lentil soup.

Juicy leaves are easy to chew, but they might be toxic

Therefore, it is no wonder herbivores turn to leaves, which are softer and juicier. Nevertheless, the herbivore's life is not carefree with respect to these leaves; although they may be chewed easily and quickly, they also may contain toxins, that is, they may defend themselves chemically rather than mechanically. An explanation of the chemistry behind this defence mechanism is beyond the scope of this book; however, some general points would be useful. Chemical substances that defend against herbivores function in three principle ways. In the mildest case, the taste of these substances is bitter, although it is important to note that what is unpleasant for humans might be a delicacy for other animals. A second, also mild, case concerns substances that reduce the nutritional value of food in various ways. Tannins, for example, bind to proteins (highly nutritional nitrogen-containing substances). The protein–tannin compound cannot be absorbed by the gut and assimilated, so a portion of the protein contained in the meal is excreted. In other words, the herbivorous animal avidly seeking protein-containing food loses part of it if the leaf it consumes contains tannins. Furthermore, saliva and gastric fluids contain enzymes (protein substances) that digest food, that is, they break down its ingredients to forms that can be assimilated by the organism. Tannins, however, bind these digestive enzymes too, striking a double blow: *neither* food proteins *nor* other food ingredients are digested. Tannins, therefore, are not toxic but reduce food absorption potential so much that the animals starve and their population dwindles. Although such animals might have eaten voraciously, their assimilation results are poor.

Why do tannins help people lose weight?

The third and most serious case involves plant tissues that contain toxins and poisons. The extent of the harm caused from consuming such substances is

Toxicity is a matter of dosage

a matter of dosage: the amount consumed within a certain period. A small dose might pass unnoticed, have no consequences, or even be beneficial by activating general defence mechanisms that can protect against other, similar toxins. A large dose, on the other hand, is a ticket to the hereafter. Fortunately, moderate doses usually cause discomfort, so the animal restricts consumption to its own and the plant's benefit.

The preceding discussion might very well lead to the conclusion that plants are not the ideal food and that their mechanical and chemical defence mechanisms restrict their consumption to acceptable limits, safeguarding some balance in nature. Although this statement basically is correct, it is not even half the truth. Therefore, a more detailed analysis of the situation is warranted. Readers, however, need not fear passages of heavy science and deep philosophy. Sometimes, science is as simple as daily life, provided one is curious and wishes to comprehend the world surrounding us.

Plants, after all, are not the most ideal food for herbivores

One might well observe that humans consume certain wild plants, some in large quantities; these plants are considered tasty, nutritious, and necessary. Do these plants have defence mechanisms; if not, why have they not become extinct? If indeed they lack chemical and mechanical defences, why have they not fallen prey to other herbivorous organisms? How is it that something humans find palatable is not consumed by other herbivores? Why do humans consume only a few tens or hundreds of the 260,000 plant species available? Strolling in the countryside, one may observe that the plants humans do not consume (because they are hard, bitter, indigestible, or poisonous) are delicacies for other animals. What is it that determines an organism's dietary preferences?

Why is a food we dislike liked by other animals (and vice versa)?

Herbivorous animals may be classified as monophagous, oligophagous, or polyphagous. For example, a human may consume some tens of plant species

The biology of food preferences

whereas the Chinese panda depends exclusively on bamboo; in other words, the panda is monophagous. At the other extreme, some locusts can consume a large number of plant species – up to 500 – although given the choice, they would prefer a more limited range. The locusts are polyphagous. Among insects, which are the top plant food consumers globally, 80% of species show some preference, that is, they limit their diet to one plant species and a few of its relatives. *Specialists*, as they are called, usually are small insects that consume herbaceous plants with young, soft leaves that are quite toxic. This is not a coincidence and points to interesting biological features involved in the selection and consumption of a specific type of food; it is a case of two-way specialisation. When an animal accepts only one plant as food, that plant is acceptable to only a few other animals. For example, some types of clover are particularly toxic because if they are damaged (which is inevitable during mastication or digestion), they produce cyanide. Cyanide is a potent poison; a few milligrams are enough to kill a human by asphyxiation. Cyanide is poisonous to the respiratory system of all organisms, both plant and animal. Such clover species, therefore, are well protected against herbivores, but how do they avoid poisoning themselves? In this specific case, the cyanide is rendered harmless because it is bound in a chemical compound called linamarin. In its nontoxic form, linamarin is stored in a special cellular organelle, the vacuole. In another cell compartment, an enzyme is stored that specialises in splitting linamarin and producing cyanide. When the compound and the corresponding enzyme are separated this way, they present no risk to the clover's respiration. If an unfortunate animal chews on this clover, the damage it causes to its leaf cells brings the enzyme into contact with its substrate, thus releasing cyanide. Therefore, the role of linamarin and the enzyme that breaks it down is defensive; they form a chemical weapon activated and

Monophagy, oligophagy, polyphagy

Why do some plants produce cyanide?

Cyanogenic glycosides: when mastication removes the pin from the hand grenade

launched when the plant is attacked. In peace time, they exist only as a defensive investment, harmless but potentially deleterious.

Natural antidotes, or why some snails are not poisoned by cyanide

Yet, clover is consumed by a snail species that, in the continuous chemical war between plants and animals, has developed an antidote, which is also an enzyme. The snail takes the cyanide and binds it to another molecule that renders it harmless. However, although this is war, both parties derive some benefit. The clover has only one predator, the snail that has managed to develop a successful cyanide detoxification mechanism, and it is not attacked by other animals. Consequently, the snail also benefits because it has no rivals so does not have to fight with other animals over the same food source. In evolutionary terms, a mutation was established in the clover population that enabled cyanogenesis, limited the plant's enemies, and increased its numbers. Balance was restored when the enemy responded with another mutation/antidote to the clover's weapon. This mutation gave it the advantage of exploiting a virgin territory not approached by other consumers.

Polysubstrate monooxygenases: enzymes for every toxin...

Many other systems of circumventing plant defences exist that are based on the capacity of some animals to counteract toxins. One of these systems is worth discussing because it seems to be an all-purpose method. The system is composed of a group of enzymes called polysubstrate monooxygenases (PSMOs). Behind this complicated name lies the following function: the system catalyses oxidation reactions by molecular oxygen, as indicated by the second half of its name. The characteristic feature, though, is that the substrate (i.e., the substance to be oxidized) is not just one but several compounds. Remember that enzymes are highly specific protein catalysts. Each enzyme receives at its active centre only one compound (substrate) with which to react, that is, the compound with which the enzyme's active centre has high affinity. Only substrates with a similar structure can "cheat" the enzyme and reach its active centre. Even then, though, the corresponding affinity is low. The specificity of enzymes is necessary for metabolism to occur in an organised manner, so that everything takes place when and where it should. In contrast, PSMOs can oxidize a multitude of toxic substances, counteracting their effect. It soon was revealed that PSMOs counteract not only natural toxic substances but artificial ones as well, which is very important in our industrialised times. PSMOs, to a large extent, are responsible for the characteristic resistance insects develop against insecticides. In other words, a noteworthy enzymic system has been selected to metabolise toxic compounds that insects take in with their food, and this system also can metabolise artificial substances insects never encountered in their evolutionary history. Thousands of artificial compounds have been manufactured and released into the environment by humans. Their toxic effect on biological systems is reflected in the hardly euphonic name given to them: *xenobiotics*. If their dose is not too high, xenobiotics are counteracted by PSMOs. A synonym for PSMO is *cytochrome* P_{450}, as this compound is an ingredient of PSMOs. This useful system is prominent in organisms that came into contact with plant toxins during their evolutionary course, that is, herbivorous animals. These organisms are most capable of counteracting xenobiotics, which explains why carnivorous animals are the most sensitive to insecticides, the use of which may sometimes have a result opposite of the one intended, that is, the friends (carnivorous insects) rather than foes (herbivorous insects) of crops are exterminated.

...causing a headache for the insecticide industry

Why are carnivorous animals more sensitive to insecticides?

Another typical property of the PSMO (or P_{450}) system is that these enzymes are induced by the products of the secondary metabolism of plants, that is, by the toxins they are to counteract. If an insect receives an artificial diet or is fed leaves with low levels of secondary metabolites, the PSMO system also is at low levels. If the insect is forced to consume leaves full of toxins, or if toxins are added to its artificial diet, its PSMO levels multiply tens of times within a few hours. If the initial dose of the toxin is mild, the insect will manage to survive. Once its PSMOs have increased, an increased toxin dose will be well tolerated. Because PSMOs attack many toxin types, a second toxin, following the first one, also will be well tolerated. If the toxin is removed from the diet, PSMOs will be reduced as well.

The foregoing discussion makes it obvious that continuous contact with low, tolerable toxin doses is beneficial in the long term, because it maintains high PSMO levels and ensures resistance to a wide range of other toxins. This obviously is true for organisms with sufficient inductive capacity and the ability to reinforce the multicollective PSMO system; humans are among such organisms. Therefore, one should not be afraid of spicy, bitter, sour, or peppery foods. No doubt, the good health of people who are in the habit of using (but not abusing) spices and a variety of plant foods is related to the induction and continuous alertness of their defence systems.

What does not kill me indeed makes me stronger (F. Nietzche), or why the mild use of toxic substances may be beneficial in the long run

To generalise, one may say that animals accept as food the plants that contain toxins for which they have an antidote. This statement answers the question asked earlier regarding what determines the dietary habits of organisms.

Let us now examine a more sophisticated system that concerns a familiar Mediterranean plant. Mammals will not approach the leaves of the oleander because they contain a series of substances called cardenolides, which are potent poisons. They also have an extremely bitter taste, so even a tiny sample will discourage a would-be consumer. However, two or three species of aphids (plant lice) and a few butterfly larvae (caterpillars) feed on these poisonous leaves. Part of these poisonous substances is counteracted by enzymes produced by the plant lice and caterpillars. Some of the remaining toxin,

Tritrophic relationships: a long way to go with oleander's poisons. Cardenolides determine which butterflies will be selected as food by insectivorous birds.

however, is stored in specific tissues, wrapped in such a way that the consumers will not poison themselves; consequently, the insects use these substances as a defence against their carnivorous predators. This is a further step in evolution; not only is the chemical defence of the oleander counteracted, but the toxic substance stolen is used by the herbivores to defend themselves against their carnivorous enemies. *Danaus* caterpillars, which feed on oleander leaves, preserve these substances even at the mature butterfly stage. When a bird attempts to eat this butterfly, its bitter taste makes the bird vomit immediately. When the bird encounters a butterfly of the same species later, it associates the colour of the wings with its past unpleasant experience and does not try to eat it again. Importantly, the bird avoids not only *Danaus* butterflies, but also other butterflies with the same wing colour, even though the latter do not have the poison. This is an obvious example of a *mimetic* phenomenon. The cardenolides of the oleander are transferred through the food chain and determine dietary relationships up to the level of insectivorous birds. They also have an indirect effect through the phenomenon of mimetism and the defence of other butterfly species that have nothing whatsoever to do with the oleander.

Cost of Armaments, Defensive Strategies, Alliances, and Nonconventional War

In the cases we examined so far, the drastic substances – bitter, astringent, or toxic – are permanent constituents of the plant cell. Given that their effect on the consumer depends on the dosage, one may infer that their deterring action is a function of their concentration. A series of interesting questions may be raised on this premise. Which plant organs are better armoured? Does defensive armour change with the age of the plant or organ; if so, what are the rules? Is there seasonality in the concentration of defensive substances in the plant? Before these questions are answered, it must be underscored that natural products come at a high cost of production. For example, the production of 1 g of phenolic (i.e., bitter and astringent) compounds costs as much as the production of amino acids, which are used by organisms to synthesise their proteins. Compared with sugar production, the cost of producing phenolic compounds is double. Terpenoids and alkaloids (which include many toxins and poisons) cost three times more than sugars and 50% more than amino acids to produce. This means that for a plant to synthesise 1 g of the toxin, it must photosynthesise three times longer than it would to synthesise the same quantity of sugars. Therefore, whatever is synthesised for the plant's defence is subtracted from its growth. A plant that must (or is programmed to) produce large quantities of defensive substances inevitably will have a lower growth rate.

Defence costs

As has been reiterated several times so far, many parameters of plant behaviour may be interpreted in economic terms, that is, supply and demand. The discussion on leaf stomata – the tiny valves on leaf surfaces that regulate the entry of carbon dioxide (to be assimilated through photosynthesis) and the inevitable exit of water vapour – stressed that the size of the valve opening at any given moment reflects the plant's dilemma: growth and development or survival? Open stomata represent a high rate of photosynthesis and growth, but also excessive water loss, leading to the risk of dehydration. In contrast, closed or semi-closed stomata lead to low photosynthetic and growth rates, but also safeguard survival. An opening is as big as necessary to achieve optimal photosynthesis under prevailing environmental conditions without risking plant survival.

Major dilemmas: growth or survival?

A similar dilemma is present with regard to the cost of manufacturing defence substances: defence or growth and development? The answer made imperative by natural selection could not but be similar to the previous one: sufficient growth/development and sufficient defence to ensure survival is not at stake.

Growth or defence?

Regarding the question of which plant organs should be best armoured, the reasonable answer is those most valuable to the plant and most vulnerable to its enemies. To analyse this issue with regard to the leaves, consider the following question: Are young leaves more or less valuable or vulnerable than mature ones? At first, the answer might seem obvious. Would any reasonable person deny that children are more valuable and vulnerable than adults, with the lowest capacity for resistance? Who can deny that one of the most important achievements of civilisation was the institutionalisation of children's protection? Who has not felt the instinctive need to help a crying child? Finally, who would dispute that it is infinitely more tragic for a child than for an adult to perish? Although an emotional approach is persuasive, in effect it is also a humanised version of necessity. In nature, there is no emotionality, only necessity, and as the beginning of this book stresses, the goal is to examine the situation from the viewpoint of plants.

Cost and value: because defence has its costs, defence investment covers the most precious and vulnerable parts

A young leaf is valuable because it must repay the plant for the cost of its construction. Its value lies in its future photosynthetic activity, which will contribute to the plant's growth, development, and survival. A young leaf does not photosynthesise intensely because its photosynthetic apparatus is still under construction. Its final surface and final number of cells have not been reached. Depending on the plant species and its developmental strategy, a leaf has a life expectancy from a few weeks in herbaceous plants to several years in the case of a conifer's needles. The longer a leaf's life expectancy, the more valuable it is; a young leaf is valuable for the future photosynthetic action it promises.

The value of a young leaf lies in its future photosynthetic promise

Young leaves are vulnerable because they inevitably are soft, meaning their mechanical defence is still minimal. A young leaf increases its size through a combination of cellular divisions and the lengthening of new cells. For this to happen, the cell wall – the exterior skeleton of the cell – must remain soft and quite elastic. The same is true for the cuticle, the hard cover of the epidermis. As a leaf grows, its vessels, the so-called nerves, multiply, lengthen, and branch out until they become more ligneous and hard, when the leaf has fully matured. If the vessels, cuticle, and walls hardened too early, this would create insurmountable mechanical obstacles for the addition of new vessels and cells. If one does not know the final size of his or her home and the final number of its rooms, it is not in his or her interest to reinforce the external and interior walls. In the same manner, the developing leaf has to remain soft until it reaches its final size.

Defensive means change with time

By observing the principle that the most valuable and vulnerable organs are the ones that must be best armoured, plants invest particularly in the chemical defence of young leaves, equipping them with high concentrations of bitter, astringent, and toxic chemical molecules. If this did not happen, the new leaf growth would be easy prey to herbivores, thus reducing the future photosynthetic capacity of the plant. It should be noted that in time, with the progressive hardening of the leaf, the investment in chemical defence stops. In other words, mature leaves often have lower concentrations of toxic molecules than younger ones. In essence, as the leaf grows, chemical defence gradually may be replaced by mechanical defence.

Short-living leaves and leaves for all seasons

A second question that may be answered in cost/benefit terms has to do with the defensive reinforcement of short- and long-living leaves. The leaves of an evergreen plant, which are programmed to live longer than a single growing season, are leaves for all seasons. Several times during their lives, they must face assault waves by herbivores. Furthermore, natural wear and tear must be considered in addition to the distress caused by animals. Particularly in regions of intense climate seasonality, leaves need to survive low winter temperatures and high summer ones, as well as periods of intense radiation and drought. The mechanical reinforcement of leaves not only is a deterrent to herbivores, it also protects leaves from abiotic stress factors. A hard leaf stands up better to wind pressure and collisions with airborne particles. Thick cell walls not only decrease nutritional value but also store water and mitigate, to a certain extent, the adverse effects of drought. Furthermore, phenolic compounds not only deter herbivores, they also protect plant tissues from harmful UVB radiation, because they absorb it before it reaches sensitive biological macromolecules, such as DNA, and causes molecular harm. Therefore, long-living leaves need to be invested with sufficient mechanical reinforcement and enriched with high concentrations of defensive chemical compounds. Their construction must be "sound," which means high building and maintenance expenses. Because of the high cost, such leaves are not replaced easily and should not be sacrificed to consumers or be left to the mercy of the environment. They are of high value. The opposite is true for short-living leaves, because they will remain on the plant for a short time only and, therefore, may be constructed more "cheaply." Their low construction cost makes such leaves dispensable and easily replaceable. Herbaceous plants and deciduous trees, therefore, have cheaply constructed leaves, disposable ones made to last for just a few months during the season favourable to the specific plant. These leaves lack strong mechanical defences or particular strength adaptations to environmental stressors; there is no point in undertaking costly fortifications for a short-term station. In contrast, evergreen plants invest highly in defending their leaves against both consumers and the natural elements; they

Leaves superficially and leaves soundly constructed

have leaves for all seasons. A typical case is that of thick, hard, leather-like leaves with thick waxy layers on their cuticles and high concentrations of secondary products in their cells, such as those of the hard-leafed evergreen trees and bushes of Mediterranean ecosystems. In this case, not only is mechanical fortification useful, but a second line of defence based on chemistry must be in place. This category of plants includes olives, carobs, lentisks, Kermes oaks, laurels, oleanders, and so on.

Why are the leaves of evergreen Mediterranean plants hard?

Therefore, the answer to the question of which plant organs are best armoured is simple: those of the highest relative value.

The second question raised earlier, whether there is seasonality in the concentration of a plant's defensive substances, requires a different analysis. As might be expected, the levels of secondary metabolites in plants usually increase during periods when they have more predators. There is no reason for an excessive investment in ammunition during times of peace. However, when the enemy is in sight, things change. Might it be that resistance is induced; if so, how is an enemy perceived?

Constitutive and induced defence

Plants have three types of defensive chemical compounds. First, there are those inherent in the plant, regardless of the threat. Second, there are compounds present in an inactive form that become toxic following an assault. An example of these is the cyanogenic compounds contained in clover, which come into contact with their catalytic enzyme once the animal has damaged the plant, thus becoming toxic. In the intact tissue, the inactive substance and the enzyme are stored in different compartments. A breach in the border between these compartments, however, results in a toxic mixture. Third, there are defensive substances that are not present before the assault but are induced by it. Particularly frequent is the *de novo* biosynthesis of substances that occurs when a plant has been attacked by microorganisms. This system is somewhat similar to the human and animal immune system: a pathogen's entry triggers reactions aimed at identifying, containing, and ultimately exterminating the invader. The immune system "remembers" the initial invasion so it will function faster and more efficiently during the next attack. The relationships between plants and pathogens are discussed elsewhere in this chapter. For now, let us focus on animal attacks.

How do plants perceive their herbivorous enemies?

An herbivorous insect's attack on a leaf can be simulated easily using a sterilised needle or scissors. Sterilisation is necessary so microorganisms may be ruled out from the experiment. Indeed, if a tobacco leaf is punctured, the level of nicotine – its main defensive compound – will increase significantly in a few days. The same is true of other plants and their respective defensive substances. Surprisingly, though, if a plant is similarly punctured by its natural enemy rather than a needle, the reaction will be multifold. Can the plant distinguish between a needle and the masticatory organ of an insect? Indeed, it can. In some cases, only the insect's saliva is needed to activate the defence system. Of the many substances contained in the saliva, only one or two enzymes are needed to trigger the response. A small puncture must be accompanied by certain molecules unique to the insect; these are detected by the plant, leading to the reaction that usually deters the invader. This means the plant is not fooled by a random prick of no consequence. It needs convincing evidence that this is not a random, transient hazard; the plant does not respond without good reason. Deterrence may be direct or indirect; for example, the increase in nicotine levels has an immediate effect because this compound is toxic. In other cases, however, the compounds produced by the plant are volatile and released into the atmosphere with no direct effect on the enemy. It is worth focusing for a while on this unusual phenomenon.

Plants continuously release into the atmosphere many volatile substances that are not perceived by humans. To a great extent, the function of these substances is unknown. Each plant has its own unique substance "footprint." Following are some interesting cases of substances whose functions are known. Recall the description in Chapter 5 of the *Ophrys* orchid releasing the pheromone of the female *Andrena* bee. The enticing compound attracts male insects to *Ophrys* flowers for the benefit of pollination. What is of interest here, however, is that the volatile compounds released after an herbivorous insect damages a leaf do not repel the insect itself. It has been found that such substances are invitations to a good meal sent to carnivorous insects, which consume the herbivorous enemies of the plant. For example, volatile compounds attract carnivorous wasps to plants assaulted by beetles, which appear to be a special treat for the wasps. In anthropocentric terms, it seems that as soon as the plant is attacked, it forms an alliance with the enemies of its enemies. It provides the latter with a ready meal and gets rid of its predators. The means of communication between plants and their potential allies are volatile substances. As in the previous case, a puncture is not enough; the herbivore's saliva also is necessary for the reaction to occur. The nature of this alliance indicates that this is not a reaction in which the prime mover is the plant – that is, a reaction with a clear and definite goal. Most likely it was the carnivores who at some point acquired the evolutionary capacity to recognise which plants are under attack, through the volatile products released by these plants. Correspondingly, the plants that acquired the ability to cry for help were favoured by this natural selection, and this quality was established because of its adaptive value.

Every plant has its own unique signature in the form of volatile substances released in the atmosphere…

…which attract the enemies of their enemies, that is, potential allies

Another example of cooperation appears to be determined by plants. Many plants are equipped with extrafloral nectaries, groups of cells that secrete nectar but are not located inside flowers; they usually are found in the shoots and stems. Therefore, nectar production in this case is not targeted to insect pollinators and is not a reward for their reproductive intermediation. To whom is it addressed, then? Nectar has a cost for the plant, and the expected payoff must be significant. Plants with extra-floral nectaries host large populations of carnivorous insects, such as the Coccinellidae (ladybirds) and ants. These insects eat the nectar and fulfill their energy needs, because the sugar content of nectar is around 20% to 60%. However, nectar is poor in nitrogen, which carnivores seek in large quantities. Because they are hungry for proteins, they will attack any herbivore daring to approach these plants. Unlike the previous case, this is not an alliance with a common target; rather, the carnivores form a kind of mercenary army, which also is kept hungry to ensure its eagerness to plunder.

Many plants use mercenaries to encounter their enemies. The reward comes in the form of nectar.

A second category of volatile substances released after damage has occurred neither repels consumers nor invites the enemies of enemies. These substances are signals to neighbours who have not yet been assaulted to inform them that enemies have invaded their territory. The adjacent plants receive the chemical signal through the air and activate their defence genes, producing secondary products to prepare for an imminent attack. This phenomenon, in which the plant-victim warns neighbours of the same species to prepare for their defence, seems altruistic. If this concept were transferred to human societies, such behaviour might seem strange. Few people who have suffered an attack would encourage their neighbours to flee and to save themselves. More likely, they would ask their neighbours to intervene, although few would do so. Therefore, what is the adaptive value of the alarm signal emitted by plants? What benefit does the victim derive if its fellow is saved? Why does the motto “every man for himself” not apply in this case? These questions are

Volatile alarm signals are received by neighbours and warn about the presence of enemies

A case of plant altruism?

difficult to answer. In human societies, altruism usually is reserved for family, much less so for strangers. Only those with special training, under exceptional circumstances – for example, soldiers at war – display altruistic behaviour towards comrades to whom they are not related. If a child fell overboard into the freezing sea, the parent would not hesitate to jump in, even at the risk of his or her own life. However, if an unknown child fell, one might look around to determine whether someone was more willing to jump in the water. Less heroic people might think of an old chest problem that might flare up. A young man might jump in to show his female partner what an affectionate and self-denying father he would be. Generally speaking, according to geneticists, altruistic behaviour among humans is as strong as the number of genes one shares with the person in danger. Perhaps, this theoretic (but reasonable) view also may be applied to the case of conversing plants. Plant neighbours are more likely than not to be related to one another. If the neighbours originated through vegetative propagation, they are clones, identical to the mother plant. If they are the offspring of sexual reproduction and seed dissemination, they have not drifted far in most cases, despite any dispersal techniques or efforts to move away from their progenitors; therefore, these neighbours are related too.

The degree of kinship among individuals of a plant population increases with shorter distances

As already mentioned, most herbivorous insects are monophagous or oligophagous, that is, they select their food from among a small number of plants and rarely make the wrong choice. How do they recognise which plant is suitable and which is not? Sometimes, these insects make their selections by trial and error. Indeed, insects may try the food, and if it is not to their taste – if it contains toxins they cannot counteract – they reject it. In that case, the animal withdraws and selects something else, and the plant is only slightly damaged. Of course, it would be more efficient for both organisms involved to determine whether the food is suitable

How do insects select plants suitable for food?

Trial and error...

before, rather than after, trying it. The former case is more advantageous, because both organisms avoid even minimal harm. Very often, the insect not only has to choose the right plant, but also must determine the most suitable developmental stage of the organ or tissue of interest. As already mentioned, the nutritional value and suitability of leaves and fruit change with the seasons and the developmental phase of the plant. Development, especially in plants, is determined by environmental signals, the perception of which informs an organism about the progress of seasons, that is, whether it is time for development or for dormancy. Then, the hormonal system is activated and turns the developmental course in one or the other direction. Therefore, one way to coordinate the activity of plants and their consumers is for both to observe the same environmental signals (photoperiod, temperature, rainfall). Another, more costly way is for insects to pick plant signals that indicate their developmental stage; these signals are olfactory, visual, or tactile. As in the case of the plant/pollinator relationship, here too the orientation of a food-searching insect is achieved through volatile substances that might be released by the plant of its choice. As the insect approaches, it uses visual signals, that is, the shade of colour and the shape of the leaf. Upon arriving at the target, the insect checks the leaf relief and the taste of its surface. Leaves differ greatly with respect to their relief and cuticle wax quality; both traits serve as reliable food quality information and help the insect decide whether to try or reject the offer. As stated earlier, *a priori* rejection based on olfactory, visual, or tactile signals saves both the organisms involved from potential trouble. If preliminary trials are positive, the insect samples the leaf interior, essentially with the aim of detecting the secondary plant metabolites. If these can be counteracted by the insect's biochemical defence systems, the feast begins; if not, the plant is rejected as food.

... or preliminary control from a distance

Olfactory, visual, and tactile signals

Insects have an excellent capacity for detecting the developmental stages of plants. For example, during their ripening, many fruits produce ethylene, a simple, volatile, low molecular weight compound that functions as a ripening hormone. If such fruits (e.g., bananas) are picked unripe and stored in a vacuum so the ethylene produced can be removed, they do not ripen. With this method (and various other tricks), unripe bananas are transported over great distances. Then, a short exposure to ethylene ripens them so they may go to market.

Volatile ethylene is a plant hormone that leads to fruit maturation...

The same is true in the case of many other fruits that can be made to ripen quickly and simply by closing them in airtight bags so the ethylene produced is not removed. Ethylene in the air, therefore, is a signal indicating that there is a ripening fruit suitable for food nearby. Correspondingly, through evolution, insects have developed the capacity to detect this volatile plant hormone through their "nose." At the molecular level, this means they have the gene codifying the olfactory receptor that binds ethylene. What is characteristic is that the molecular nature of this insect receptor is the same as that of the respective ethylene receptor on fruit cells. When the ethylene produced by unripe fruit is bound by the corresponding receptors of fruit cells, this triggers the ripening process. When bound by the insect olfactory receptor, it tells the insect that there is a ripe fruit in the vicinity. Therefore, the plant and the insect that consumes its fruit probably have the same gene for producing the same receptors, which, however, have different physiologic results. Is this a case of convergent evolution or gene stealing?

...while also indicating to insects that fruit is ready to be consumed

Ethylene receptors in plant fruits and insect olfactory organs are the same. Therefore, the corresponding genes are the same. Is this a case of convergent evolution or gene stealing?

Insects need plants not only for food, but also as a surface on which to deposit their eggs. They do not lay their eggs just anywhere but carefully select the plant tissue on which to deposit them. As stated previously, the caterpillars hatching from the eggs need abundant and suitable food. Many interesting interactions have been observed in this field, based on the chemistry of secondary plant products. Insects think the same way humans do. Who does not want the best future possible for their offspring? Some examples follow.

Why do insects perform a hormonal analysis of leaves before depositing their eggs?

Just as some insects have learned to measure the fruit-ripening hormone, others have learned to measure hormones called gibberellins, which are related to leaf development. A large quantity of gibberellins on a leaf indicate that it will be alive for quite some time. Reduced quantities indicate that the leaf soon will age. Some female locusts interested in depositing their eggs on cabbage can determine its developmental stage by measuring its gibberellin levels. If they are not high enough, the locusts preserve their virginity. If the levels are satisfactory, the locusts' genitalia mature and they release into the atmosphere the pheromone inviting the males to intercourse. Soon the eggs are deposited on the cabbage, and the young locusts about to hatch have a guaranteed food supply. In other cases, the pheromone attractive to the opposite sex is a compound the insects take from the same plant, modify chemically, and release into the atmosphere.

Hormonal games with the enemy

The crowd of wooers, however, creates problems for the plant. Besides dallying and flirting with the females, the males also need sustenance. Therefore, the significant increase in the population of herbivorous insects does not favour the plant, because its survival is now threatened.

When the plant lice population living as parasites on plants becomes a nuisance, many of these plants produce the hormone of insect migration

No doubt, the foregoing discussion has helped the reader conclude that the plant will try to discourage the invaders chemically to reduce their population by imposing or supporting their escape. The following case demonstrates how the plant achieves this goal. Plant lice (aphids) come in two forms: one with wings and one without. When the plant lice eggs hatch, nonwinged forms develop, propagate on the leaf, and colonise it. When the colony becomes too large and overexploitation of the leaf reduces its nutritional value, the aphids realize this, grow wings, and fly to another plant.

The developmental change from nonwinged to winged forms is associated with the so-called insect juvenile hormone; when its levels are high, insects remain wingless. However, in the continuous chemical war between plants and their consumers, some plants have managed to control the numbers of plant lice before they get out of hand. These plants produce substances that induce an earlier conversion to the winged stage. In other words, the plants interfere with the insects' hormonal system and force them to leave.

Hormones are harmful if they appear at the wrong time in the wrong dosage

This system is quite successful and has been adopted by many plants. Hormonal developmental regulation is finely tuned in an organism; it is based on producing the right hormone in the right quantity at the right time. If any of these three prerequisites are not met, developmental irregularities of tragic consequences result. As a case in point, many conifer and pteridophyte (fern) tissues produce and accumulate ecdysone, the hormone necessary for insect metamorphosis. These metamorphoses require *ecdysis*, the moulting of the old external skeleton of the insect and the creation of a new one, which gives the insect a totally different form. Ecdysone is produced at the exact time the insect's previous developmental stage ends. Ecdysone has no physiologic function in plants; therefore, one may assume it should not be produced by them. However, not only is it produced by certain plants, it is produced at very high concentrations. All an insect needs to do is prick a plant producing this hormone, and it becomes laden, at the wrong time, with a quantity of hormone several times greater than that which would lead to early moulting. Therefore, insects avoid feeding off the tissues of these plants and if they do eat them, they suffer developmental irregularities, sterility, and early death.

Other plants produce animal oestrogens (female hormones), which at first sight are useless. Plant hormones have a chemical structure totally different from that of animal hormones. The presence of oestrogens was discovered by coincidence in tulip bulbs. In the Netherlands during World War II, people had no choice but to consume such bulbs, and irregularities were noticed in the monthly ovulation cycle of women, as well as disturbing reproductive dysfunctions in men. What could be the reason some plants produce oestrogens? Certain clovers can produce isoflavonoids with a structure similar to that of female hormones in quails. When rainfall is abundant and clover grows without a hitch, its isoflavonoid levels are low and quails consume it without any problem. In other words, when times are good, clover can give up some of its biomass as food for the birds. However, when there is a drought and clover growth and development are problematic, it cannot afford such a loss. Therefore, isoflavonoid levels rise and quails end up having reproductive problems. Their egg numbers are reduced and so, consequently, is the number of adult birds. In other words, reduced clover growth during a drought is reflected in a corresponding reduction in the quail population so that clover survival is not at risk. This population control is based on the production of oestrogens by plants.

Oestrogens produced by clovers control quail populations

Therefore, there is a nonstop arms race in the context of a merciless chemical war between plants and herbivorous animals, whereby every improvement in the arsenal of one leads to an improvement in the arsenal of the other. The outcome of this struggle is balance in nature: the vegetation is not totally consumed, yet no animal starves either. To a great extent, this wonderful natural balance is based on the plants' rich biochemical arsenal. However, defence is costly, and defensive expenses are undertaken at the expense of growth and development, which is why plants try to reduce expenditures by allying with the enemies of herbivores.

Balance in nature is underpinned by an arms race, chemical warfare, and alliance building

Herbivorous animals are not the only enemies of plants. Like animals, plants also are attacked by microorganisms, many of which are pathogens. However, they also are attacked by other microorganisms, mainly fungi, whereupon the relationship they develop is mutually beneficial and turns into *symbiosis*. Some typical cases of symbiosis and their biological significance are discussed in Chapter 8.

There are many similarities – as well as differences – in the way plants react to animal assaults and attacks by pathogens. Mechanical defences such as thorns, hard leaves, crystals, and a thick cuticle are not very effective against such microorganisms. Therefore, defence is organised mainly on a chemical basis. A distinction also is made here between permanent and induced immunity. The former is supported by chemical substances always present in plant cells and inhibits an invader's growth. The latter is provoked by the invasion and results in the activation or *de novo* synthesis of defensive biomolecules.

The sequence of events is as follows. The aim of a microbe is to use the plant as food. For the microbe to enter the plant, it must break through the walls and membranes of plant cells and feed on cellular juice. With regard to the part of the plant aboveground, the only way an invader can approach is in the resistant form of spores, which are airborne or transported by animal carriers. The atmosphere is quite dry, and these organisms can travel only in the form of spores. Their landing on a plant surface does not necessarily mean the pathogens can grow. For a spore to germinate, the landing surface must remain moist for an adequate period so the spore has a chance to sprout. Herein lies the first obstacle: most plant surfaces have on their epidermis - particularly that of their leaves – tiny hairs comprising one to several cells. In fact, these hairs are appendages of epidermal cells with a secretory function. They are alive and arrayed relatively sparsely on the plant surface. They should not be confused with the rich trichome made up of dead, air-filled hairs, which are visible to the naked eye as intensely reflecting surfaces on the leaves of certain plants. The role of the

Why does wet weather favour the attack of plants by pathogenic microorganisms?

Defence at the borders

latter is to reflect part of the solar radiation so it is not absorbed by the leaves, leading to overheating. The trichome also limits water loss in leaves and obstructs small insects from entering the nutritious leaf interior. In contrast, the small living hairs are visible only under a microscope and, in effect, are exocrine glands. In other words, they have very high metabolic activity, producing special substances they excrete to the leaf surface, just as sweat glands do in animals.

Living secretory hairs as the first line of defence against an invader

What kinds of substances are excreted, and what is their role? Usually, they are secondary metabolism products that, as already mentioned, are substances related to plant defence. Therefore, this is the first conclusion: it is not only the cellular interior that is armoured with repellant or toxic substances, but also the surface of the organs. A second generalisation has to do with potential recipients of such substances: if these substances address neighbouring plants, that is, if they inhibit their neighbours' growth, thus improving the competitive potential of the donor, then the substances are water soluble, that is, they are washed by the rain and enter the soil. If, however, the substances are involved in local defence, they must remain fixed on the surface of the organ and not be washed away. Therefore, the chemical structure of these substances makes them hydrophobic and insoluble in water. They accumulate in a layer on the surface of plant organs and comprise their first line of defence. They act as antimicrobial agents, inhibiting the germination of bacterial and fungal spores and limiting their growth if germination does manage to take place. From a chemical point of view, these substances are terpenoids or hydrophobic flavonoids. The second group has an extra significant function: they highly absorb harmful UVB radiation, preventing it from reaching the cell interior, where it might affect vital biological macromolecules such as DNA, causing mutations. In other words, besides their antimicrobial actions, these substances also serve as a sunscreen.

The dual action of surface flavonoids: antimicrobial and sunscreen agents

Recently it was proven that secretory hairs on leaves also excrete a series of relatively hydrophobic proteins (i.e., proteins that cannot be washed away by water easily) with antimicrobial action. This finding is particularly important because similar proteins are also secreted on the skin of animals.

Surface defensive substances are a preventative. The skin of an organism is the border separating it from the external world, and naturally it is preferable for an organism to strengthen its borders and fend off enemies before they can cross. Many microbe spores land on plant surfaces, but very few break through the first line of defence. As long as a city's walls stand up to an attack, its peaceful operations need not change. However, when the front collapses, the martial strategy changes from passive/preventive to active to thwart the enemy's advance. New mechanisms must be activated, and there is general mobilization. Peaceful occupations continue only to the extent that they support defensive efforts, whereas other, purely defensive operations take precedence. The organism's internal chemical industry modifies its products, turning its production line to the manufacture of ammunition to biologically and chemically bombard the enemy. If the assault worsens, the organ retreats, leaving behind a "scorched earth." It is a suicidal sacrifice by the part to save the whole.

When the front collapses, defensive tactics change

Peaceful and war activities

Following spore germination, microbe cells divide and create various cellular hyphae, that is, they branch out into cellular nets searching for a way in, such as an incidental rupture in the plant epidermis, damaged or weak points, or, in the case of leaves, natural gas exchange "channels" (i.e., the stomata). After the microorganism enters the leaf's internal free space, its next target is to break into the plant cells. Its weapons usually are enzymes that break down plant cell walls. Of course, the plant has to respond the same way; this is how a chemical war starts: the plant synthesizes and

Regarding wild plants, if an enemy penetrates their walls, even if the chemical war is inexorable, the parties involved survive, suffering only minor damage

releases chemical substances to neutralize or render inert the weapons used by the attacker. It is a merciless chemical war in which the quality of the armament and the speed with which it is implemented by the two adversaries decide the final outcome. Of course, in the case of wild plants, the plant hardly ever succumbs. The battle's outcome usually involves limited damage: the plant survives whereas the microorganism is not fully exterminated. The microorganism's growth and development are contained, but it almost always has fed sufficiently and leaves the war zone after constructing several new spores that can try their luck elsewhere. The case of cultivated plants – whose attack must be dealt with by humans using fungicides and other medicines, otherwise the whole crop might be destroyed – is not a typical example. Humans have selected cultivated plants through cross-breeding and by making various choices to ensure high yields of ultimately useful products, but these plants have lost most of their ability to resist attack. Cultivated plants are like lap dogs – they can grow only with human support. They cannot compete with wild plants, nor do they have effective resistance against pathogens and herbivorous animals.

The plant perceives the invader by recognizing its chemical weapons

What is it that activates a plant's defensive mechanisms? How does it know it is being attacked?

Pathogen recognition has a molecular basis. A plant's defence is stimulated by various molecules produced by the invaders to break down the plant's resistance. For example, the enzymes a fungus produces to break through the wall of a plant cell may be used as a stimulant. Alternatively, the molecular composition of the invader itself might provide sound and useful information. The structure of the cell wall of the assaulting fungus reveals its identity. Finally, the very fragments of the wall of the plant cell itself, produced during the fungal attempt to penetrate it, are a clear indication of the attack. These processes trigger the onset of plant gene expression, which was silent until then, resulting in the biosynthesis of a series of defensive proteins. There are many lines of defence, including counter-attacks. To counterattack, the plant produces mild antibiotics called phytoalexins, the aim of which is to contain the growth of the invading organism. Phytoalexins originate from the biosynthetic pathway of phenylpropanoids, a true biochemical factory with a huge capacity for producing a multitude of biomolecules. Therefore, besides mild antibiotics, the same pathway also leads to the production of structural blocks directed at the point of assault and strengthening plant resistance at the cell wall level. The initial response is to spray the enemy with any antibiotic at hand to delay its advance while repairs are made and the walls are strengthened at the points where the enemy launched its main attack.

Wall repairs and counterattack with antibiotics

For the aforementioned events to occur, organization is necessary, along with messengers to transmit the right signals to the right place at the right time. Leaving details aside, two important messenger molecules are salicylic acid and nitrogen monoxide. The former, in the acetylated form the pharmaceutical industry produces, is the well-known antipyretic agent known as aspirin. It is a messenger molecule only for plant cells, and when transmitted to neighbouring cells it is a signal for them to reinforce their defence. Remember the involvement of salicylic acid as a signal in the induction of thermogenic respiration in malodorous flowers (Chapter 5). As for the latter messenger, nitrogen monoxide also is produced in animals, in which it plays an important role in regulating membrane permeability, DNA transcription, and erectile function.

Chemical messengers describe the situation at the front. One of them has a structure similar to that of aspirin.

One result of the synthesis of these signals is the so-called oxidative burst, during which plant cells turn oxygen into free toxic radicals, which are particularly active in the deconstruction of cellular membranes.

The oxidative burst caused by plants at the points of attack is suicidal in nature. It has an adverse effect not only on the invader, but on the defender as well.

If all previous lines of defence collapse, the technique of scorched earth is adopted: cells adjacent to the point of invasion commit suicide so they may not be used as supplies by the enemy

If things worsen further, the plant activates its last line of defence – the hypersensitivity reaction – which brings about the quick death and deconstruction of plant cells attacked by the pathogen. Therefore, when the pathogen finally manages to demolish the walls, there is nothing to find in the castle. The enemy is trapped at the point of assault, which may be observed as spots, usually brown, on the leaf. The leaf cedes some ground yet contains the invader within it while managing to maintain its occupation over the rest of its area. Plant cell death is very fast, organized, and programmed, somewhat resembling a suicide, so the defenders do not fall into the enemy's hands. Usually, this programmed cellular death also includes a group of adjacent, nonassaulted cells. It is as though the plant retreats while burning the crops and blowing up the bridges to block the enemy's advance.

To conclude, resistance against pathogen microbes starts with the molecular recognition of the invader and continues with the implementation of a defensive plan aimed at discouraging the attacker. The walls are reinforced and the enemy is bombarded with antibiotics. If that fails, more toxic substances are produced (oxidative burst) that affect not only pathogens, but plant cells as well. Finally, if this method also is unsuccessful and the enemy persists, the whole territory under attack – along with its adjacent regions – is evacuated. This allows time for the rest of the tissue to better organize its defence and to entrap the pathogen in an area where, no longer having anything on which to feed, it creates resistant forms of dissemination (spores) and departs.

In other words, the pathogen's attack and the plant's resistance response are similar to chemical warfare, using tactics and strategies encountered in human conflicts as well. Furthermore, as shown in previous pages, volatile chemical substances play a major role in communication among animals as well as between plants and animals. The next chapter shows that certain substances produced by plants and some micro-organisms in the soil are dissolved in groundwater and serve as signals for recognizing and establishing symbiotic relationships. Therefore, whether as defensive weapons or in mutually beneficial relationships of every possible combination, the substances an organism produces and disseminates into the environment comprise its identity. These substances are not released randomly, and they play an important role in establishing relationships with other organisms in the vicinity. However, one possible combination is missing from those mentioned earlier: the plant–plant pair. This raises a question: Do plants communicate chemically with one another; if so, why?

How do Plants Communicate with One Another?

So far, two cases of plants releasing volatile substances have been mentioned: plants that advertise a reward for pollinators and leaves that are harmed by herbivorous insects. In the latter case, the odour emitted attracts carnivorous insects. Might the recipient of the volatile substance be another plant; if so, what is the purpose?

Competition among plants: the role of growth rate and of substances released into the environment

Plants recognize one another's existence to secure optimum access to light, water, and mineral nutrients from the soil. The levers in this race for dominance are growth rate and the consequent strategy used to exploit reserves. Characteristics such as germination season (which differs from plant to plant), the tendency to creep or climb, an annual or a perennial life cycle, seasonal growth arrest, geotropism, phototropism, and many others are adaptive traits and make up a plant's competitive potential. Plants also compete with their neighbours through chemical means. This phenomenon is called *allelopathy* and the substances involved *allelopathics.* If this rationale for plant interaction is taken a step further to include not only competitive but also facilitative relationships (because these exist as well), then such substances are referred to by the general term *allelochemicals.*

Volatile substances and substances washed to the ground by rain

The substances involved in these relationships, usually phenols and terpenoids, are dispersed into the environment in two ways: they are either released into the atmosphere in a volatile form or washed into the soil by rain. As mentioned earlier, the gaseous environment around each plant is characterized by the presence of many substances characteristic of its identity. However, the same wealth of nonvolatile substances may be discovered by collecting rainwater dripping from leaves and running down the plant stem. These substances comprise losses for the plant; it might have prevented their leakage and preserved them for its growth and development. Therefore, it is reasonable to assume there is a role these substances play, a function they perform, so losses are paid off with some potential gain.

The case of the walnut tree

The phenomenon of allelopathy has long been known, based on the observations of farmers and floriculturists who noticed that certain plant species thrive when grown in proximity, whereas others prefer isolation and solitude. A well-studied case is that of the walnut tree, under the crown of which many plants fail to grow; something prevents their seeds from germinating. The leaves, stems, and branches of the walnut tree produce a substance (1,4,5-trihydroxy-naphthelene glucoside) that is washed to the ground by rain. Naphthalene derivatives are quite toxic, but by bonding to glucose, their toxic action is neutralized. In other words, naphthalene glucoside is not harmful to the walnut tree, because the tree produces its nontoxic form. Only when the fuse, that is, the glucose, is removed does toxicity appear. This is a very common trick among plants. Recall that cyanogenic glycosides (defensive substances in many plants, from clover to bitter almond) are harmless until the glucose is removed from the compound, whereupon the potent respiratory poison called cyanide is released. The fuse is removed by enzymes (glycosidases), which are quite common. All that is needed is for an animal to bite into a plant, destroying the physiologic separation between the glycosidase and the cyanogenic glycoside, for the poison to be produced. Under normal circumstances, the poison is appropriately stored and isolated, as one normally (should) do with toxic waste.

With regard to the walnut tree, rain washes the potentially toxic compound, but the plant does nothing to prevent such loss; rather, it assists it, as the rest of its useful intracellular substances remain inside when it rains. In other words, the loss is selective. Many plants secrete toxic substances from their epidermal glands to the leaf surface, where they remain until the next rain. In the soil, the compound released by the walnut tree is hydrolyzed by bacterial enzymes (bacteria are omnipresent), and the toxic 5-hydroxynaphthoquinone that is derived inhibits, even at infinitesimal concentration levels, the germination of other plant seeds.

The walnut tree marks its territory the same way dogs do

The sequence of events is as follows: The toxic substance is washed to the soil only when it rains; therefore, it exercises its inhibitory action exactly when the soil – because of the rain – is sufficiently wet to favour seed germination. What would be the point in disseminating a toxic substance into dry soil? Therefore, the timing is economically sound. The plant exploits a natural phenomenon – rain – to suppress the populations of other competitive plant species within its territory. In some way, it marks its vital space, which is the same as its crown projection surface area. However, one may sometimes observe plants growing under a walnut tree, but the reason for this should be obvious:

Plants exploit rain not only to take in water but also to contain the growth of their rivals the plants growing there are those that, in the course of their evolution, have managed to "circumvent" the toxic effect by producing antidotes.

Other plants use volatile allelopathics with a great affinity for soil colloid particles. Others secrete readily soluble toxins from their roots directly into the soil. The action of allelopathics, in cases in which this has been clarified, is to limit the growth of rootlets and make them lose their geotropic orientation, to inhibit the development of root hairs, and to limit cellular respiration and division.

Is this a common phenomenon in nature? This is where things get rather difficult. Soil, where allelopathics end up, is a complex system with rich bacterial metabolic activity that soon modifies the chemical nature of the toxin, sometimes rendering it inactive. Furthermore, soil can neutralize toxins through their adsorption by soil colloids. Therefore, the donor plant must secrete significant quantities of the substance for an effective concentration level to be maintained in the soil. It seems, however, that it is worth the plant's trouble to do so, and some such systems have been described quite convincingly.

Chemical competition among plants is more intense where resources are scarce

Ecosystems in which plant competition must be chemically regulated are those in which the growth of plants is restricted by the scarcity of a significant factor. Many desert plants seem to deploy chemical marking of their vital space; visually, this becomes apparent because no clusters are created. Desert plants usually are loners; they cannot afford to compete for the most precious resource of their habitat: water. An exception is the case of the so-called phreatophytes, the roots of which are deep enough to reach the underground aquifer. Indeed, phreatophytes seem to seek the company of other plants; part of the water they draw up they spill around the vicinity, thus facilitating the establishment of a "community" of plants. This way, perhaps, each phreatophyte reduces its likelihood of being

Socialites and loners

overconsumed by herbivores, now that a wider choice of plant species is available. This phenomenon has already been described in the discussion on how plants operate like environmental engineers (Chapter 3).

Although less stressful than desert ecosystems, Mediterranean ecosystems – with their long, warm, and dry summers – provide several examples of chemical interactions. The characteristic bare zones around some Mediterranean bushes have been attributed to allelopathic phenomena.

Chemical competition and chemical facilitation: how plants select their neighbours

In some cases, a donor plant not only inhibits germination of alien seeds in its vicinity, it also facilitates germination of select seeds of other species. It is as though plants choose their company. A special branch of plant biology called plant sociology is aimed at studying the structure of these typical accumulations and the reasons behind their establishment. Indeed, in many cases, if a plant sociologist is given the name of the dominant plant in an accumulation, he or she can predict – within reason, of course – which other plants might be encountered in the system. At first, one may reasonably assume these plants would be those best suited to the specific habitat, soil, and climatic conditions. However, it appears the plants themselves, through chemically modifying their surroundings, may express their preferences and determine, to a certain extent, which plants to prevent and which ones to foster in their vicinity.

The discussion of plant defence concludes with a description of a noteworthy phenomenon related to the recipients of volatile substances produced by leaves after being assaulted by either microorganisms or insects. It already was mentioned that some plants produce volatile substances that attract the carnivorous enemies of herbivorous insects, resulting in a reduction in the latter's population. In other words, the recipients of this olfactory chemical signal are carnivorous insects. It is a kind of help call, like a cat being called in to eat mice or a ladybird to eat plant lice. The plant invites its enemy's enemy, who – objectively – is its friend. Recently, however, it has been suggested that in some cases the recipients of the volatile signal from the plants under attack may be neighbouring plants of the same or a different species. As a result, the neighbouring plants increase the levels of their chemical defence before they are even approached. Therefore, in the eventuality of an attack, which is quite likely because their neighbour has already been assaulted, these plants have gained some time and are ready to face the enemy from a better position. If this phenomenon is examined from a teleologic perspective, it might be considered an alarm signal warning about an imminent threat. This is not an audible siren but a chemical one. Defence mechanisms in the recipient plant are activated not in response to an actual threat, but in response to a well-founded suspicion.

When chemical sirens "sound"

During the best-organized experiment (related experiments are particularly difficult), plants of a wild tobacco variety were planted at various distances from plants of a sage species, the leaves of which scientists had been clipping every so often to imitate the action of herbivores. As a response to this "damage," sage leaves developed an induced chemical defence. Paradoxically, the intact tobacco leaves responded by synthesizing toxic alkaloids, provided that one prerequisite condition was fulfilled: a location within a 15-cm radius of the "damaged" sage plant. If the latter was intact, the tobacco did not synthesise alkaloids no matter how close it was. The reasonable hypothesis for such communication between two plants, which is restored following damage to either of them, as well as the result of such communication (the intact plant is prepared for the eventuality of assault), has led to a range of literature under the sensational title "talking plants."

Plants talking to each other?

The chemical basis for such communication is as follows: Plants that have been harmed produce a phenolic substance called jasmonic acid. This substance, which is not volatile, is used as a messenger molecule to transport the message of attack within the plant assaulted, so remote tissues may prepare for a potential attack. Part of the jasmonic acid, however, undergoes methylation (i.e., a $-CH_3$ group is added to its molecule) to produce methyl-jasmonate, which has different chemical properties. The new molecule is not soluble in water and is volatile under normal temperature conditions. In other words, it can easily escape into the atmosphere, operating as a signal for adjacent plants to start strengthening their defence.

Is this the development of a chemical language of communication among plants? This is a provocative view that cannot be easily resolved yet. As with all environmental signals, a definitive judgment requires discovery of the perceptive mechanism. How does the plant "smell" this volatile organic compound? What is its molecular receptor? How is the signal amplified internally after it has been perceived? Although all these questions remain unanswered, it is not impossible for such a language to indeed exist, as has already been clarified in the case of animals.

Although the benefit to the recipient is apparent, the adaptive value of releasing the signal for the donor is, at least initially, difficult to explain. Plants continuously and mercilessly compete with one another, and natural selection might not favour the provision of useful information to rivals. On the contrary, it might be preferable to keep the assault somewhat secret so it may silently spread to neighbours while the victim may somehow find relief. Furthermore, it would be more consistent, from a competitive point of view, for the plant to turn its enemy's attention to its neighbours, so the enemy might leave the damaged plant alone. Attributing this phenomenon to altruism, which is favoured by natural selection if it involves relatives (so common genes are preserved in the population), is not applicable, given that the release of methyl-jasmonate favours not only

Could plants be eavesdropping on the troubles of their neighbours?

individuals of the same species but those of other species as well. Because no other biologically reasonable hypothesis exists, for now it may be assumed that the release of this drastic substance by the prey is biochemically inevitable and that intact plants have, in the course of their evolution, acquired the capacity to use it to their own benefit. This is like developing the ability to eavesdrop on one's neighbours. This scientific field has only just begun to develop and is expected to bear fruit in the next decades.

Chapter 8
Symbioses Galore

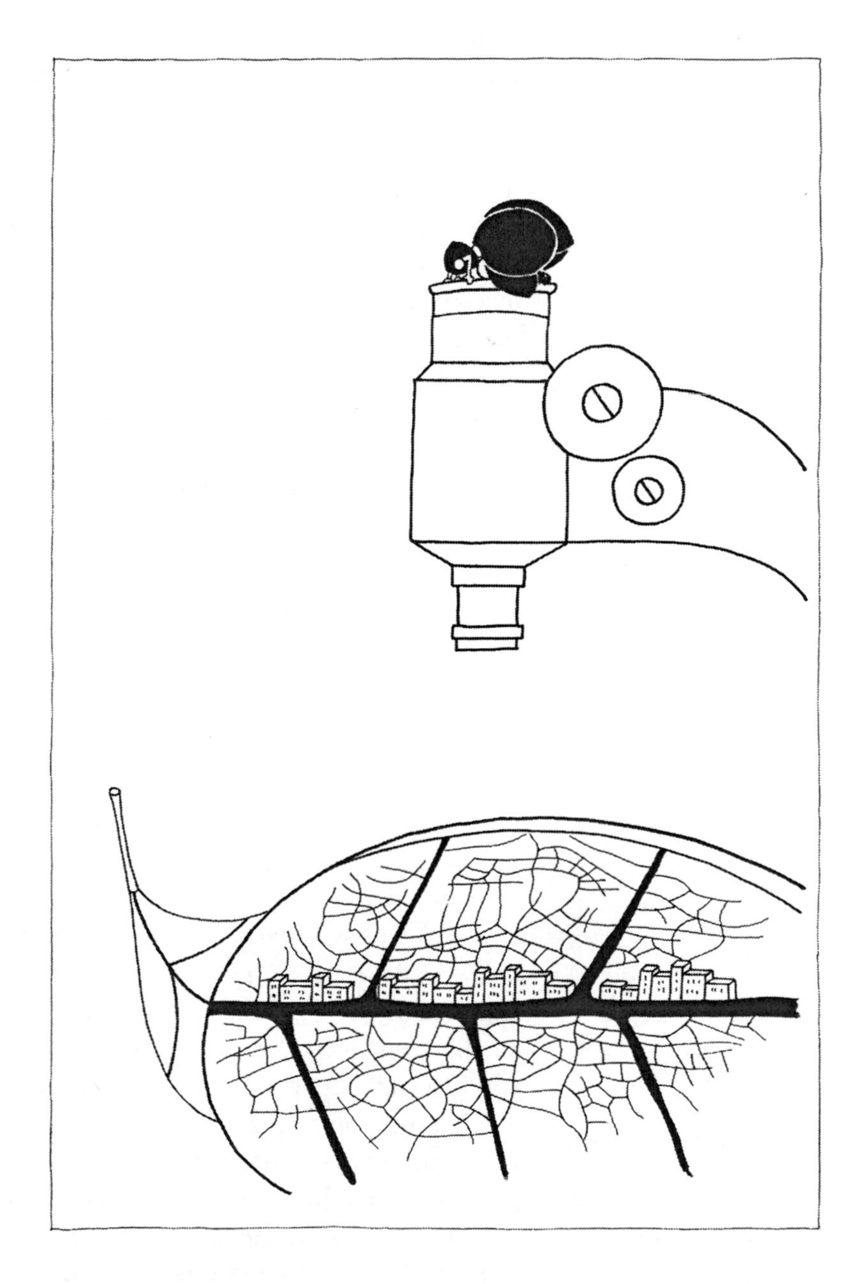

The Fine Balance between Symbiosis and Parasitism

Previous pages included a discussion on the relationships between plants and pathogenic microorganisms as well as herbivorous animals. These relationships have some potential benefits for one of the two parties involved (the predator), provided of course that they manage to break the prey's resistance. The prey, on the other hand, suffers some damage, the scale of which depends on the outcome and duration of the war, the scale and quality of the arms used, and the persistence of the adversaries involved. The predator depends on the prey, which provides it with food; in other words, the predator is helpless without the prey: the prey produces and the predator tries – more or less successfully – to appropriate the prey's capacities, that is, its nutritional self-sufficiency and biosynthetic wealth, to complement its own inadequacies. The most successful strategy for the predator is to take what it needs without exhausting or debilitating the prey, without risking its extinction. If the host is exhausted or killed, another one has to be found. However, plants as prey have proven hard nuts to crack. Successful resistance and mild parasitism are two sides of the same coin, the golden rule of balance, which keeps vegetation almost intact yet does not let herbivores go hungry.

Mild parasitism, or why predators should not exhaust their prey

Also discussed earlier are plant–plant interactions that are allelopathic. The release of allelochemicals by plants into the environment inhibits the growth of neighbouring plants, yet it is somehow harmful to both the donor and the recipient of the substance. The production of the allelopathic substance comes at a cost to the former, who – if there was no threat – could invest in growth and development rather than defence. Similarly, it comes at a cost for the latter, which must either invest in neutralizing the inhibiting substance or tolerate its adverse impact. This is an antagonistic relationship.

Fierce antagonism and honest reciprocity are two extremes of a behavioural continuum

At the other end of the spectrum of interactive relationships between organisms – at the antipodes of competition, where everyone loses out – there is reciprocity, a relationship in which the involved parties are not rivals – attackers or defenders, predators or prey – but partners cohabiting in a mutually beneficial manner. However, reciprocity comes at a cost too, because it requires concessions and compromises. Above all, it requires the partners to observe strict rules of the game; if these rules are violated by either party, the relationship immediately lapses from reciprocal to antagonistic or parasitic.

Profit and loss in symbiosis

Between the two extremes of antagonism (which harms both parties involved) and reciprocity (in which both parties benefit), there is a continuum of relationships involving various levels of profit and loss for both partners. Therefore, besides parasitism (in which one benefits and the other loses), there is commensalism, in which one benefits and the other is not affected, and amensalism, in which one is not affected and the other is harmed. The borders between these categories are unclear in the continuum of possible profit and loss levels.

All these relationship types are included in the general concept of symbiosis between different species, a symbiosis that entails cohabitation and food sharing. The large number of species compels them to coexist; this coexistence sometimes turns into fierce competition and sometimes into mild and productive cooperation, with all sorts of situations of controlled tolerance in between.

The strategy of mild symbiosis in nature is more common than supporters of "the struggle for existence" would like

Today, there are many more examples of mutually beneficial symbiosis than biologists would have imagined a few decades ago. The Darwinian view of evolution guided by brutal competition and dominance of the stronger or best "adapted" left little room for the possibility that the forces of evolution and natural selection might well be moderate. The endosymbiotic theory (the major significance of which is highlighted in Chapter 4) explains the presence of certain organelles within plant and animal cells following the invagination (without digestion) of one bacterium by another, when each bacterium possessed a different biochemical armament and a different physiology and used different ways to exploit environmental reserves. Within the new cell that emerged, the two formerly different organisms cohabited, complementing each other. The complex was favoured by natural selection and proclaimed a new, more advantageous organism. Since then, numerous attempts at cooperation involving various species – not only bacteria but also multicellular organisms – have proven successful enough to become established. Recall the example of the fig tree and the tiny wasp (see Chapter 5), an exceptionally close symbiosis – so close in fact that neither party could reproduce if the other disappeared.

Through symbiosis, organisms gain access to resources they could not acquire by their own devices

Similar to the example of the endosymbiotic episode of bacteria, symbioses generally may be viewed as attempts to acquire qualities one lacks through the recruitment and appropriate manipulation of organisms that have these qualities. There is an initial trial stage that may look like a paid labour arrangement or, in the worst case, like captivity. If the trial is successful, the next step is to persuade the captive (or guest) to remain forever, voluntarily or involuntarily. The host may offer rewards so that it would not be in the guest's interests to leave the relationship; the host may increase the reward (recall the relationship between plants and their pollinators that was established through continuous improvement of the quality of flower rewards), or the host may concede total ownership while keeping the usufruct. If the initial relationship tends towards captivity, the host artfully removes some of the guest's capabilities so the guest realises it is not in its interests to go back to being independent.

Just like a betrothal before a wedding, in symbiosis there is an initial trial period for the two organisms

Let us look at the relationship between humans and dogs. Dogs descended from wolves and were tamed by humans some 2,000 to 3,000 years before sheep and goats, around 11,000 BC. Taming the dogs did not improve human nutrition directly; dog meat is not edible. If humans had the olfactory capacity of dogs (i.e., former wolves), they would not need the animals – they could discover game on their own. If humans had not lost their body hair, they would not have gained much by sleeping with dogs on cold nights at the end of the recent Ice Age. Indeed, it seems one of the first uses primitive humans had for dogs was that of a blanket. Through this taming, humans indirectly improved their hunting capacity, borrowing a packet of qualities they themselves did not possess, a packet of genes codifying good olfactory receptors and a warm coat. Humans entered a new world, that of perceiving and mapping the environment through odours, a world they exploited without ever truly experiencing it through their own senses.

Through taming dogs (symbiosis), humans improved their own capacity to locate game without developing an excellent sense of smell themselves

Symbiosis was established when dogs ceased to be wild. By taking a share of the game they discovered without having to fight it (this part was undertaken by humans, who were more skilled at it), dogs lost part of their potential for independent survival. A return to freedom was not favoured; a dog attempting it would have to compete with the better-equipped wolf. The inevitable outcome would be extinction or assimilation. Although stray dogs survive in packs today, it is only because there are not enough wolves against which to compete. The human–dog pairing was favourable and ultimately selected; the symbiosis was successful. *Taming*, in anthropocentric terms, is synonymous with a meek and soft character, and the control of wild instincts. For the dog, it means concession of privileges, dependence, and an attitude of servile obedience. Once the dog learned to stand on its hind legs, it had lost the game of independent survival. The same is true

Symbiosis might entail significant concessions, even total relinquishment of one's previous habits

in the case of chloroplasts, which no longer can escape from plant cells, because a portion of their vital genes has been transferred to the plant cell nucleus.

Too Close a Symbiosis Leads to Deformation

How much can an organism change when involved in a symbiotic relationship? Although dogs changed in character, they still look like wolves. The history of this symbiosis is recent, a mere 13,000 years, and it is a loose relationship at that. However, what about long-term relationships with symbioses established millions of years ago, which by their nature had to be particularly close? How unrecognisable does an organism become as a result?

A fig that does not expose its genitalia (flowers) and the miserable life of a male wasp that never sees the light of day

As shown in Chapter 5, plants of the *Ficus* genus are exempt from the rule of apparent flowers. Instead of advertising the presence of flowers with intense colours and a conspicuous position as other plants do, the *Ficus* hides its tiny flowers in a closed inflorescence that can be entered only by those who hold its key. These flowers never see the light of day. The demand for strict exclusivity in this symbiotic relationship made the fig inflorescence look like a fruit and demoted the poor male wasp to a miserable, pitiful, blind nit. In every other aspect, a fig tree is a normal plant. The only concession caused by this state of affairs has been the concealment of its flowers. How much can symbiosis deform a plant, though?

How disfiguring can symbiosis be?

Some plants exist as parasites on other plants. In this symbiotic relationship, one gains and the other loses. Parasitic plants do not grow in the soil but on other plants, specifically on permanent organs of the host plant. There is no point in their growing on leaves, flowers, or temporary organs, which soon will be shed. Therefore, they germinate on the stem or root of the host

Rootless parasitic plants

and feed on its juices. These parasites do not need to develop roots with suitable adaptations for effectively absorbing (usually scarce) soil nutrients and (often insufficient) water. Parasites find everything ready-made from host vessels, so they have no need for a well–branched-out root or for rootlets. The same is true for the root cap, the tissue that perceives the gravitational direction and turns the root towards the earth's centre. Furthermore, parasites need not worry about mistakenly taking up any toxic inorganic elements. This is the responsibility of the host, which, through its roots, makes sure only suitable elements enter. In other words, the parasite is not exposed to toxic elements because it receives inspected and certified food. The parasitic plant's only requirement is a blunt stem in contact with the host's vessels, an intensely modified root called a *haustorium*. Some parasites, such as mistletoe (*Viscum album*), which usually lives on the branches of conifers, direct their haustoria towards xylem vessels and receive water and minerals for their own benefit at the expense of their host. Others, such as the *Cytinus ruber*, a parasite living on the roots of Mediterranean bushes of the Cistaceae family, attack phloem vessels, from which they receive not only water and minerals but also organic substances such as sugars and amino acids. In other words, they enjoy a full diet and have no need to photosynthesise because they exploit the host's photosynthetic capacity. In plants such as *Cytinus*, mutations resulting in the loss of leaves and stems are not unfavourable. These parasites do not have to photosynthesise, so they need not emerge above the ground surface, because the source of their food is the host's root. Their body is made up only of the haustorium. They remain obscure underground and become apparent only during their reproductive season, when their flower-bearing stem emerges temporarily. Their life is similar to that of fungi, the *hyphae* of which live continuously underground and make only a

Rootless and leafless parasitic plants, that is, heterotrophic plants

brief appearance at the surface – as toadstools and mushrooms – during the reproductive season. Unlike other plants, parasitic ones of this kind are not autotrophic but heterotrophic, as are animals and fungi.

An extreme morphologic concession due to parasitism is found in the plant named *Tristerix aphyllus*. Its seeds are disseminated by birds, and they germinate when they land on certain cactus species. Their young, fine roots penetrate the cactus through the stomata or cracks in the epidermis and settle in the form of a web among the green photosynthetic cells of the host. Nothing on the outside of the cactus indicates that it is hosting a parasitic plant in its interior, and nothing inside the plant indicates that this web of non–chlorophyll-containing cells belongs to another plant. The parasite, at this stage of its life, has no root, stem, or leaves; it has no organs or tissues and consists only of a web of undifferentiated cells feeding off the cactus cells. The parasite becomes a normal plant only during the flowering season, when it produces a stem, penetrates the cactus epidermis, and emerges, producing leaves and flowers, which will remain until the seeds ripen and are disseminated completely. During this brief span, the cactus bears leaves and flowers, which of course are not its own. Through common experience, it is known that cacti have no leaves, because these have turned into thorns. Cacti photosynthesise through their thick green stems, which are also a water-storage organ.

Cell webs that turn into plants only during the reproductive season

Why do some cacti bear another plant's flowers?

The most extreme case of parasitism, and a quite sensational one, is that of *Rafflesia*, a parasite on the trunk of the climbing tropical plants of the Vitaceae family (which includes ordinary grapevines). *Rafflesia* is a very rare species that lives in tropical forests of southeast Asia. This so-called plant lives as a web of cells inside the host's stem and never develops a root, a stem, or leaves. It feeds parasitically from the host, but without causing it great damage. The only organ that gives away its presence as a plant is its flower, and

Rafflesia: a plant with a bodiless, huge, and revolting (from an olfactory point of view) genital organ. Not recommended for use in a vase or as a gift for your date.

what a flower it is! This genus holds the world record for the flower with the largest diameter; in the most impressive case – that of *Rafflesia arnoldii* – it may be as large as 1 m. The flower can weigh up to 10 kilos and is not suitable for a vase, not so much because of its size but because of its smell. The flower gives off a strong, particularly repugnant smell, like that of a body in an advanced stage of decomposition. As in the case of the faeces-like smell released by the flowers of some plants of the Araceae family, the odour that is so hideous to humans is a stimulating advertisement to scavenger insects of the presence of protein-rich food. For the setup to be more convincing, the *Rafflesia* flowers with their malodorous emissions are also fleshy, usually red with black or brown spots, intensely reminiscent of a neglected slaughtered animal, at least with regard to colour and smell. Despite the successful staging, the play does not satisfy the scavenger insects, which depart indignantly because they have found no flesh; however, they have already been sprinkled with pollen, which they will transport to other *Rafflesia* flowers so that pollination may take place.

Ultimately, the only thing parasitic plants are not prepared to lose is their genitalia

The Nitrogen Problem on Planet Earth and a Plant–Bacteria Symbiosis of Huge Impact

Not all symbioses are parasitic, though. Many plants are involved in a series of symbiotic relationships beneficial to both partners. A typical characteristic of such symbioses is that the organisms involved are totally different from each other; they belong to different kingdoms, with different nutritional demands and products, so that cohabitation and cooperation pay off. The relationship between plants of the *Ficus* genus (which includes the fig tree) and the tiny wasps that spend almost their entire life within the reversed

Cohabitation of plants and microbes

inflorescence of the plant (figs) has already been described. This close cohabitation makes it possible for both partners to reproduce and perpetuate themselves, whereas neither would have been able to do so on its own. In two other important examples of symbiosis, the plant's "dinner guests" are microorganisms. These relationships, beyond their biological significance, are also of great economic importance, given that they improve the growth and development of the plants involved.

Why is nitrogen necessary?

One of the most necessary elements for all organisms is nitrogen. It is an ingredient of amino acids (which make up proteins), organic bases (comprising DNA and RNA), chlorophyll (which is necessary for photosynthesis), and of a series of other vital organic biomolecules. Where do organisms find nitrogen, and how do they assimilate it? With regard to most other inorganic nutrients, it can safely be claimed that their origin lies in the layers of the earth's crust, within which they form various chemical compounds with one another. Rock detrition leads to the dissolution of these compounds so that their elements enter groundwater, ocean water, or other water bodies. Organisms, mainly plants, receive the elements from these locations, and the nutrients then follow the food chain: plants → herbivores → carnivores → decomposers (mainly bacteria and fungi that decompose dead organisms) → soil → plants, and so on. However, although all organisms need large quantities of nitrogen, it is very scarce in the earth's crust and its release from the rocks by weathering is extremely slow. Practically the only source of nitrogen for organisms, then, is the atmosphere. Almost all the nitrogen found in the soil today – in its oxidized form as nitrate and nitrite anions (NO_3^-, NO_2^-) or in its reduced form as ammonia (NH_4^+) – originated from the atmosphere and was produced through one of the following three processes:

"Ashes to ashes and dust to dust" is valid for all minerals necessary for organisms, except nitrogen

All the nitrogen found in organisms and the soil today ultimately originated from the atmosphere

In descending order of significance, atmospheric nitrogen fixation today is of biological (bacteria), industrial, or natural (lightning) origin

- Natural processes, mainly electrical discharges during storms. Locally high temperatures and pressure levels developed during a lightning strike catalyse nitrogen oxidation by atmospheric oxygen; the nitrogen turns into nitrogen oxides, which are dissolved in rainwater and fall to the ground. Natural processes are responsible for 1 % of atmospheric nitrogen fixation.
- Biological processes, that is, fixation by the so-called nitrogen-fixing bacteria, which live either independently in the soil or symbiotically with certain plants. This biological process, which is discussed in detail later, is currently responsible for 80 % of atmospheric nitrogen fixation. Before the industrial and technologic revolutions, this figure was almost 99 %.
- Human activity. The need to fertilise crops to ensure higher yields, combined with technologic progress and new ways to manufacture nitrogenous fertilisers using atmospheric nitrogen, has gradually increased the human contribution to atmospheric nitrogen fixation. In principle, this is useful for crop productivity but gradually has led to excessive fertilisation, that is, eutrophic problems in adjacent water masses. Surplus nitrates cannot be adsorbed by soil particles in great quantities, so they are washed away and excessively fertilise lakes in the vicinity, altering their physiologic balance. Excessive growth of phytoplankton near the surface of the water mass and the consequent increase in the population of organisms feeding on it cause two undesirable effects. First, the intensity of light reaching the bottom layers is reduced; therefore, the photosynthesis of sedentary, benthic algae decreases, resulting in a local reduction of O_2 levels. Second, the growth of populations in the top layers inevitably leads to an increase in dead organisms deposited at the bottom. Their decomposition by bacteria there consumes O_2 and creates intensely anaerobic conditions, resulting in further

Excessive nitrogen fertilisation destroys aerobic life in closed water basins destruction of aerobic life forms. This phenomenon is called *eutrophism*.

After this brief description of the nitrogen cycle, let us return to nitrogen-fixing microorganisms. These specific microorganisms live independently in the soil and might be aerobic or anaerobic, photosynthesising or non-photosynthesising. They are the only organisms capable of fixing atmospheric nitrogen. Consequently, and only with regard to nitrogen, they are at the base of the food chain, because almost all the nitrogen needed by other organisms, in effect, has been fixed by these bacteria. *Almost all* refers to the entire history of life on our planet, from its dawn to about the beginning of the last century, when humans managed to chemically manufacture nitrogenous fertilisers. Today, almost half the nitrogen necessary for cultivated land comes from biological fixation; the rest is provided by chemical fertilisers. In contrast, in natural ecosystems, the contribution of human-fixed nitrogen is negligible, except in ecosystems in direct or indirect geographic contact with cultivated land.

The only organisms that have the privilege of atmospheric nitrogen fixation are certain bacteria

Such bacteria can fix nitrogen because of the presence of nitrogenase, the enzyme that catalyses the reduction of atmospheric nitrogen to the level of ammonia. The reaction requires high energy consumption: to reduce one nitrogen molecule, 16 molecules of ATP and eight molecules of reduced ferredoxin are necessary. For these molecules to be manufactured, great quantities of sugars also must be consumed (oxidised). The ammonia produced can then be used for the biosynthesis of amino acids, proteins, nucleic acids (DNA, RNA), chlorophylls and other nitrogen-containing organic molecules. These molecules, primarily produced by nitrogen-fixing bacteria, are released into the soil after they die and decompose. They are then used as a nitrogen source by all other organisms. As a result of the decomposition of nitrogen-containing organic

Atmospheric nitrogen fixation carries a high price

macromolecules, the soil contains mainly three forms of inorganic nitrogen: a reduced form (NH_4^+, ammonium cation) and two oxidised forms (NO_2^- and NO_3^-, nitrite and nitrate anions, respectively). The ratio of reduced to oxidised forms depends on the oxidising–reducing capacity of the soil, which in turn is determined by oxygen concentration and the presence of aerobic or anaerobic bacteria that use the inorganic nitrogen forms as electron donors during respiration.

Through their roots, plants can absorb all three available inorganic nitrogen forms, which they can process and use to synthesise all the other nitrogen-containing organic compounds they need. Animals lack this capacity, because they need processed organic nitrogen, at least in the form of certain amino acids. In other words, animals are heterotrophic with regard to nitrogen. In a narrow sense, however, plants are not absolutely "autotrophic" regarding nitrogen because they cannot take it up from the atmosphere directly; rather, they need the intermediation of nitrogen-fixing bacteria.

Some plants host nitrogen-fixing bacteria in their roots. They provide them with room and board (sugars) and receive assimilated nitrogen (ammonia) as a reward.

Although nitrogen-fixing bacteria are everywhere, it is better to have a device to produce ammonia at home rather than relying on the uncertain presence of such bacteria in the vicinity. Therefore, some plants enrich their food content in fixed nitrogen by establishing symbioses with nitrogen-fixing bacteria in their roots. The roots involved are deformed and present nodules. The bacterium provides ammonia and the plant provides energy (in the form of sugars) to the ammonia-manufacturing factory (i.e., the bacterium). This symbiosis is exceptionally successful for two reasons: First, plants are completely self-sufficient with regard to sugar because of photosynthesis; they usually have a surplus of sugar they easily can afford to give away, whereas nitrogen quantities usually are insufficient. Second, symbiosis forces the bacteria to dedicate themselves exclusively to producing ammonia. When these bacteria live in the soil independently, they fix only the

A nodule is an exceptionally efficient microscopic factory manufacturing nitrogen compounds

nitrogen they need. When they are involved in symbiosis, however, they are urged to specialise, that is, to intensify ammonia production to meet the plant's needs as well. This is achieved because the bacteria are not compelled to find the necessary energy resources to fix atmospheric nitrogen, because that energy is now provided by the plant. This symbiotic system, therefore, is quite effective with regard to nitrogen fixation, something empirically known to farmers once they recognised that soil became more productive when they rotated crops from time to time by cultivating a legume species between periods. Legumes (the Fabaceae family, which includes cultivated beans, clover, broad beans, lentils, chickpeas, peas, and numerous wild, noncultivated species) are capable of symbiosis with nitrogen-fixing bacteria of the *Rhizobium* genus.

This is why legumes fertilise the soil

Besides legumes, 20 other angiosperm plant genera from other families also have this capacity. In all cases, the symbiosis is established in the plant's roots. Besides these plants, there are a few representatives of two age-old plant groups, the pteridophytes (ferns) and cycadophytes (relatives of conifers), that have developed the same capacity. A noteworthy case is that of *Azolla*, a tiny pteridophyte (with a body a few centimeters long) that floats on stagnant freshwater. Inside its tiny leaves, among the photosynthesising cells, are filaments of cells of the cyanobacterium *Anabaena*, which is capable of fixing atmospheric nitrogen. The *Azolla/Anabaena* system is particularly useful in rice fields, where it enriches the water with nitrogen, thus increasing rice productivity. Rice does not establish similar symbioses and relies on inorganic nitrogen forms it takes up from the soil. If the waters that flood the rice fields contain no *Azolla*, then farmers ensure they are artificially enriched with this organism. The particularity of the *Azolla/Anabaena* system is that the symbiosis is established in the leaves not in the roots, as in the case of legumes.

Tiny floating factories of the Azolla (fern)/ Anabaena (cyanobacteria) company manufacture fertiliser for rice fields

Naturally, the symbiosis of legumes with rhizobia attracts interest because these plants are used as human food, and the mechanisms behind the symbiosis are known to a certain extent, facilitating the comprehension of such an important biological phenomenon. Various questions may be raised, such as, How does the plant recognise the bacterium, and vice versa? Recognition is important given that the symbiosis is very specific: Every plant species uses a specific bacterial species. How does the bacterium enter the root? Where does it settle? What structural changes are necessary at the root to prepare the infrastructure, the home in which the bacterium will live?

How do legumes recognise their future bacterial partners?

The whole recognition process is a good example of a conversation between the plant and the bacterium using a chemical and genetic language. The plant secretes from its roots specific substances (flavonoids and betaines) that are recognised only by the appropriate bacteria. These substances are chemotactic, like the flower smells that attract insects. The bacterium actively approaches the source of these substances, the root. The same substances awaken in the bacterium specific genes that produce other substances secreted into the environment; these in turn are recognised by the plant, and chemical contact is established. When the bacterial substances are received by the cells of a root hair, genes of this hair are activated and cause the hair to bend so it may trap the rhizobia. Bacterial enzymes cause the selective decomposition of root hair cell walls and create a hole through which the bacteria penetrate the root's interior. When the bacteria reach suitable positions in the cortex, plant cells are stimulated through a new genetic conversation and proceed to membrane synthesis; bacteria enter these membranes and start dividing. These microscopic pouches are the final locations at which the bacteria establish themselves. Cell divisions at the root lead to formation of the so-called nodules, masses of plant cells

Negotiations for setting up a company use a chemical and genetic language

containing the bacteria. The peripheral cells of the nodule turn red because of the presence of legume haemoglobin, a protein similar to mammal haemoglobin. It is the only case of plants synthesising haemoglobin; its role is to bind oxygen (as in mammals) to maintain an extremely low oxygen concentration within the nodule. It is imperative that oxygen not reach the "factory" of atmospheric nitrogen fixation because the enzymic mechanism of fixation is sensitive to oxygen and does not function smoothly in its presence. In short, legume haemoglobin serves as an antioxidant shield. At the same time, angiogenesis is induced in the region, that is, the network of roads carrying energy (sugars from leaf photosynthesis) to the factory is created, which also removes from the nodules the useful products of atmospheric nitrogen binding in the form of amino acids. Nodule membranes, which incorporate entire bacterial cell populations, are equipped with the correct entry and exit gates and are in appropriate contact with adjacent vessels so that substances may flow in and out smoothly. Through such symbiosis, plants enriched in bacteria may indirectly increase their independence by assimilating nitrogen from an inexhaustible atmospheric source, paying for it with part of their sugar surplus, that is, reduced carbon.

Why do legumes synthesise haemoglobin?

Nodules are a regular factory importing unprocessed nitrogen and energy and exporting processed organic nitrogen

Fungi and Plants Connected in an Underground Web

Mycorrhizae: plant and fungi cooperatives

Another useful and popular symbiosis in which many plants participate is the so-called mycorrhiza. As in the previous case, cohabitation occurs in the root, yet there are some differences. The partner here is not a bacterium but a fungus. Fungi are microorganisms known for creating *mycelia*, webs of cells in contact with one another. In mycorrhizae, the web is spread out and in close contact with the root's interior cells but also extends outside the root over great distances within the soil. This is a mutually beneficial symbiosis in which the plant provides the fungus with carbon in the form of sugars (taken from its surplus, as in the symbiosis of legumes with rhizobia), whereas the plant takes water and minerals, mainly phosphate and sulphur oxides, from the fungus. In essence, this symbiosis allows the plant to expand much further than the limits of its roots. The fungus absorbs water and minerals through its fine hyphae and gives part of them to the plant roots; for these services, it is rewarded with sugars.

Mycorrhizae facilitate the uptake of water and inorganic mineral salts

Plant and fungus symbioses are quite popular; 80 % of plants examined to date are involved in them. They are also specific, that is, a limited number of fungus species can establish symbioses with a limited number of plant species. This specificity presupposes recognition between the two partners, that is, a chemical and genetic conversation. It seems, however, that mycorrhiza symbiosis is important for adequate catering not only to individual plants but to plant societies in general. It has been proven that forest trees are interconnected through their mycorrhizal systems into an extensive forest web. In other words, the hyphae of the fungi in symbiotic contact with a particular tree branch out, expand, and, like a wire network, form symbioses with the roots of all neighbouring trees. Although invisible to an observer on the ground, a subterranean web of fungal hyphae connects trees to one another. The operation of this connection is not yet fully understood, nor is it known whether trees exchange information through it. What is certain is that they exchange carbon in the form of sugars, which might be of particular significance for young saplings in the understorey, which do not receive much light and therefore have a low rate of atmospheric

They also interconnect plants through an underground web…

carbon dioxide assimilation through photosynthesis. It might be a case of altruism to benefit less-favoured neighbours. Besides, younger neighbours most likely are the offspring of older trees in the vicinity, so this is a case of parental care, in the broadest sense of the term. All organisms, even those believed to be unaware of their own existence (unlike humans), are interested in and care about the good health of their descendants, the gene packets they have disseminated.

... and provide younger neighbours with carbon, in a demonstration of altruism and parental care

Remember that underground webs are also established in cases of plants that can extend their roots over long distances and grow new shoots from these underground rhizomes, without cutting off communication with the mother plant. Whole forests might be made up of such clone-trees. Mycorrhizal webs seem to establish communication among trees that do not have vegetative (nonsexual) reproduction through cloning. In such cases, underground communication might be particularly significant.

Plants leading a parasitic life on the mycorrhizal web

The mycorrhizal web is exploited by some strange parasitic plants that establish underground contact with the fungus and remove part of the sugars the fungus has received from neighbouring trees as a reward for its services. This is a reversal of the common tree/fungus relationship: usually, the tree provides carbon and the fungus provides water and minerals. In this case, the parasitic plant has no chlorophylls and cannot photosynthesise; it leads its parasitic life not on the roots or stem of another plant, but on the hyphae of a fungus, intercepting the fungus's reward or, in other words, serving as an indirect parasite on the tree in symbiosis with the fungus. These parasitic plants have rudimentary roots in contact with the fungus and emerge aboveground only when they are in bloom.

Are there symbioses with fungi in plant organs aboveground? During the past 15 years, it was discovered that certain members of the Poaceae family (formerly Gramineae, or grasses, which include several cultivated species, such as wheat, maize, and sugar-cane) feed fungal hyphae within their leaves. The technical term for such fungi is *endophytes*. They seem to be contributing to the defence of leaves against herbivorous animals through the synthesis of toxic or deterrent substances. As a reward, they receive room and board from the plants – in the form of sugars.

Some acacias feed bodyguard-ants...

Are there other mutually beneficial and exclusive relationships between plants and insects besides that of the fig tree and the wasp? Indeed, there are. Tropical acacias in Central America have voluminous hollow thorns, within which the plants host aggressive ants. The ants feed on the secretions of extrafloral nectaries, the nectar-producing secretory cells not contained inside the flowers. Nectaries lie in the stem and branches of the plant. Providing food to the ants is the only reason the extrafloral nectaries exist, and providing a home for the insects is the reason the thorns are hollow. Once again, this is a case of providing room and board. In exchange, the ants undertake the role of bodyguard for the acacia, defending its territory and attacking any living being that approaches, whether it be an insect, a mammal, or even another plant. If a neighbouring plant dares raise its height and cast its shadow on the acacia, the ants attack it and destroy all its leaves. This symbiotic relationship is obligatory: in the absence of its bodyguards, the acacia becomes easy prey for herbivores because it does not have any noteworthy means of chemical defence in its leaves. Furthermore, this species of bodyguard ants has not been found living anywhere except inside the acacia thorns.

...and laurels let out chambers to mites

If the underside of a Mediterranean laurel (*Laurus nobilis*) leaf is examined carefully under a magnifying glass, tiny bumps forming a small enclosed space with a small entrance may be observed at the points where the central "nerve" crosses lateral ones. The technical name for these enclosures is *domatia* (small chambers), a term that signifies exactly what they are: within them reside tiny mites. There is no satisfactory explanation as to why laurel leaves (and leaves of other plants) grow such formations during their development. Most likely, the mites provide some service to the laurel in exchange for room and protection. However, the domatia on laurel leaves should not prevent one from using bay leaves in cooking lentil soup – not because the mites are tiny and invisible to the naked eye, but because they moved in search of a new abode as soon as the leaf was cut.

Animals that Photosynthesise, or an Evolutionary Episode in the Making

Animals, of course, have not been any less successful in establishing symbioses with other organisms, as shown in the few examples presented so far in which one of the partners is an animal. Describing symbioses of animals with other animals or microorganisms is beyond the scope of this book. However, there are two cases worth mentioning in which the other party is a photosynthesising organism – not a plant, but an alga.

How corals photosynthesise

Corals indisputably are animal organisms; nevertheless, they share two characteristic features with plants: they are immotile (i.e., they do not move) and they photosynthesise. Immobility adds a further similarity: just like plants, corals are capable of vegetative reproduction through cloning. Masses of tens or hundreds of individuals on a coral reef are made up of genetically identical, cloned individuals with a common ancestor. The capacity for photosynthesis is acquired through symbiosis. Within their bodies, they host single-celled algae called zooxanthellae. These algae are captured at the first life stages of a coral and are imprisoned in subcutaneous sites where they photosynthesise. This is why corals that pursue symbiotic relationships limit themselves to shallow waters, in which there is ample light for the tenant to photosynthesise. This nutritional contribution is complementary: the tenant provides a dietary supplement to the coral, making the latter less dependent on external food sources. Therefore, the coral becomes less heterotrophic without losing its capacity to capture tiny animals and plants approaching within the action range of its small tentacles. All the coral has done is to incorporate a light-collecting system in the form of an alga, a gadget photosynthesising for its benefit.

In the symbioses examined so far, the participants are two different yet complete organisms. However much evolution might have led to the loss of physiologic functions or characteristic morphologic features of the organisms involved, they still look somewhat like their distant, nonsymbiotic ancestors. The parasitic *Rafflesia* plant has lost its root, stem, and leaves, but its flower is still a plant flower. The marine worm *Olavius algarvensis* enters a symbiosis with an internal garden of many and various bacteria with a characteristic task distribution: one group digests the worm's food, another uses the first group's residues as food, and the third undertakes the processing of the worm's waste. The worm does not need (and no longer possesses) a digestive or excretory system. It maintains its circulatory, muscular, and nervous systems, but in all other respects it essentially is a sac full of bacteria, without which it does not exist as an organism. There is, however, another characteristic symbiosis case, in which the host does not accommodate the whole symbiotic organism but only part of it – in effect, only one of its cellular organelles – while it devours the rest. This is the green photosynthesising slug of the *Elysia chlorotica* species, which lives in marine water.

There are many green animals whose colour is not the result of chlorophyll; other chemical and physical structures exist with these visual characteristics. However, the green colour of *Elysia chlorotica* is indeed

Elysia chlorotica: a photosynthesising slug or a motile alga?

caused by chlorophyll. An initial, superficial explanation might attribute this slug's colour to its food: indeed, it feeds on algae. However, this explanation does not hold water: whatever an organism eats is digested and decomposes within its alimentary canal into small molecules, which the organism has the biochemical capacity to use as an energy source or as raw materials for synthesising its own characteristic molecules. In the case of chlorophyll, in particular, its deconstruction is necessary for one more reason. The molecular structure of chlorophyll is such that it absorbs light and, stimulated by it, transfers one electron from its molecule to a suitable adjacent molecule, thus initiating a flow of electrons within chloroplast membranes, resulting in the synthesis of energy-rich biomolecules. If the chlorophyll is removed from its strategic position in the chloroplast membranes (which happens when the chloroplast is deconstructed during digestion) – in other words, if it is not bound to the appropriate adjacent molecules – it becomes toxic. This is because when free chlorophyll absorbs light, it transfers the excitation energy to oxygen (which is omnipresent), turning it into a free oxygen radical. This radical is particularly toxic, with a great capacity for destroying cellular membranes and vital biological macromolecules, including chlorophyll itself. Therefore, it is absolutely necessary for chlorophyll to be decomposed during digestion into smaller molecules that do not have such undesirable properties.

The elysia slug has chloroplasts

An anatomic observation of slug cells under a microscope proves that its green colour is the result of whole, intact chloroplasts from the alga the slug consumes. This raises the first question: Besides the chloroplast, the alga cells contain all other ordinary organelles, such as nuclei, mitochondria, and ribosomes. As expected, all these are digested and deconstructed within the slug's alimentary canal, while paradoxically the chloroplasts are exempted from this process and remain intact. It is not yet known how this exemption is achieved. Even more impressive (although expected after the foregoing discussion) is the position the intact chloroplasts finally occupy: they are distributed in relatively superficial cell layers of the slug, locations that receive light and do not lie in the deep darkness of its interior. Naturally, one suspects a symbiosis that renders the slug photosynthetic, that is, the theft of a device that is then positioned appropriately to achieve high performance levels, the same way solar water heaters are placed on terraces. It is easy to prove that the slug, when lit, assimilates carbon dioxide from the atmosphere and turns it into sugars, just as a leaf does. Indeed, it is a photosynthesising slug, or in other words, a mobile leaf. It can survive throughout its life with no other source of carbon, provided it can receive the necessary minerals from its environment; it has become autotrophic, just like a plant.

Just like plants, it "eats" carbon dioxide from the atmosphere...

One may remonstrate that this is not an unprecedented phenomenon. In previous pages, it was mentioned that corals host single-celled algae (zooxanthellae) for the same purposes. In the case of the elysia slug, however, there are a few innovations. It is not just a matter of colour; elysia slugs are dark green like a mature leaf, whereas corals usually are a pale yellow-green. The corals' colour is a result of the small concentration of chlorophyll in the zooxanthellae and their small population within the corals, which, however, is a minor difference. What is most important is that elysia slugs move, whereas corals, just like plants, are immotile, fixed forever to the point at which they were established. This probably is the first case in the history of life on Earth of a multicellular, motile photosynthetic organism. The term *multicellular* is stressed because many single-celled algae have flagella they use as a propulsive propeller: the flagella move but cannot carry the organism very far. The slug, albeit not a champion long-distance runner, can intentionally select a position of optimal light

... and it is the first multicellular, motile, photosynthesising organism

intensity to avoid too much or too little light. It already was mentioned that plants also seek optimal positioning for light by modifying the direction of their growth or by moving their leaves accordingly, always within the context of an immotile existence. It definitely is a sign of progress that the search for the optimal position is assisted by an animal-like movement, albeit a crawling one. Moreover, besides being carbon dioxide absorbers and sugar synthesis devices, chloroplasts are also oxygen producers. The elysia slug can potentially exploit and explore anaerobic environments, in which it has a competitive advantage over organisms that are similar to it but have not adopted chloroplasts.

The adoption of chloroplasts rather than a whole photosynthetic cell is the second major innovation. In the symbioses mentioned so far, intact organisms rather than their organelles were involved. What is the problem? What is the disadvantage of adopting part of an organism? The problem concerns the maintenance of the organelle.

A strange CV

Let us start at the beginning and examine the elysia slug's curriculum vitae. The young slug is not green; it does not inherit the chloroplasts from its parents, as plants do. Regarding plants, the huge number of chloroplasts they contain in their adult life originate from a minimal number of plastids (organelles that are chloroplast precursors), which are contained in the ovum. In other words, they inherit the chloroplasts from their mother's side. Plastids keep dividing and are distributed into dividing cells as well. If the plastids are located in areas in the plant that receive even a little light, they differentiate into chloroplasts. Every new generation of elysia slugs, however, must acquire its own chloroplasts *de novo* through its food. Although the slugs can feed on various algae, they keep the chloroplasts of only one species, called *Vaucheria litorea*. If a slug is deprived of the opportunity to eat *V. litorea*, it remains colourless and heterotrophic.

It seems only *Vaucheria* chloroplasts are suitable for the elysia slug, or are the only ones it has developed a trick to maintain. Once the chloroplasts are established, they are preserved throughout the lifespan of the slug, even if it does not eat *V. litorea* again. The life expectancy of the elysia slug is around 9 months. Towards the end of its life, the slug reproduces but does not transfer the chloroplasts to its offspring.

The elysia slug does not inherit chloroplasts from its parents. It seizes them from algae and keeps them active for the rest of its life.

The preservation of algal chloroplasts within elysia slug cells for 9 months surprised plant biologists, at least those who had conducted laboratory experiments with isolated chloroplasts. When chloroplasts are removed from leaf or alga cells, they survive in vitro for a few days at the most. This is because they are semiautonomous organelles, that is, their maintenance requires genetic information, only part of which is contained in their DNA. The rest of the information is contained in the nuclear DNA. In simple terms, the inevitable turnover of chloroplast proteins and their necessary resynthesis requires cooperation with the cell nucleus. Usually, part of each chloroplast protein is codified by chloroplast genes and is synthesized within it. The remaining part is codified by the nucleus and synthesised in the cytoplasm; it enters the chloroplast through special gates to suitably join the rest of the protein part. How, then, are *Vaucheria* chloroplasts maintained for 9 months in the slug cells once the nucleus of *Vaucheria* cells and the corresponding DNA have been digested? Studies have shown that the remaining information regarding the chloroplast's preservation has been deposited in the DNA of the host's nucleus. This event must have occurred during the first attempts to establish this symbiosis, at some point in the distant past. In other words, the slug made sure that, along with the photosynthetic gadget, it kept the specific toolkit for its repair.

It also has the tools (genes) to repair them when the need arises

Horizontal gene transfer: a neglected evolutionary force

The technical term for this phenomenon is *horizontal gene transfer*, as opposed to *vertical transfer* from ancestors to descendants. Horizontal transfer is rare among superior organisms but quite common among bacteria, in which cells belonging to different species establish cellular contact and exchange genes (i.e., new qualities), which might at some point prove useful. Stealing and keeping the tools for repairing the chloroplasts of the alga *V. litorea* was absolutely necessary for the elysia slug so that a lifelong relationship (or imprisonment) might be established. Readers are referred to Chapter 4 for a description of the primary endosymbiotic phenomenon that occurred during the initial evolutionary stages of life on the planet and contributed to the creation of eukaryotic cells through the imprisonment of previously independent bacteria; these ended up in cellular organelles, which today are called *mitochondria*, *chloroplasts*, and *flagella*. Finally, it should be mentioned that among biologists, there is a view gaining ground that evolution is based not only on mutations but on horizontal gene transfer from organism to organism as well. Mutations are random changes that appear in a single gene; horizontal gene transfers are loans of entire gene populations. Both phenomena are subject to the judgment of natural selection and are established as new qualities, provided they give the organism competitive advantages. From this point of view, the elysia slug has in its nucleus a packet of genes inherited from its distant ancestors, a second packet it once removed from the nucleus of the alga *V. litorea* (i.e., the toolkit for its chloroplast repair) and installed in its own nucleus, and a third noninherited packet every elysia slug generation must acquire again during the first stages of its life by eating *V. litorea*. This is the DNA of the alga chloroplasts.

Mutation, as an evolutionary force, concerns only a single gene at a time. Horizontal transfer concerns packets of gene populations.

Chapter 9
Deviations from the Basic Biological Type

Carnivorous Plants

When the principle of being autotrophic is violated: Carnivorous plants...

A fundamental biological principle, namely that plants are autotrophic and photosynthesising whereas animals are heterotrophic, seems to have been violated in several of the examples presented in previous pages. Some examples show plants that have lost their photosynthesising capacity living as parasites of other plants or fungi; others show animals with an acquired photosynthesising capacity, albeit on loan. Some plants have deviated even further: they have become carnivorous. However, one should not think of blood, terrifying canine teeth, sharp claws, and an imposing appearance. Carnivorous plants are small; most are obscure tiny herbs whereas a few might reach 50 cm in height. As for the animals consumed, they are insects no bigger than a fly. It would be more accurate to call such plants insectivores rather than carnivores.

...are, in effect, small, obscure insectivores

The first accurate observations on these strange plants were made by a particularly restless and visionary intellectual figure of the 19th century – Charles Darwin – the very same person who proposed the theory of evolution based on natural selection. Darwin postulated that the bodies of dead insects gradually decomposing inside the "traps" of carnivorous plants probably provide the latter with some useful substances. With the poor analytic tools of his time, all Darwin could collect was indirect evidence. However, he was astute enough to provide the traps of such plants with small pieces of meat or cheese and to observe that the plants grew better and seemed healthier compared with others that had not received such a meal. In other words, he conducted an experiment that was simple but respected all the rules of experimental ethics.

Insect traps: they immobilise, kill, and digest the prey caught; they then absorb nutrients

A plant's trap is usually a modified leaf that is green, contains chlorophyll, and photosynthesises normally. Carnivorous plants have not given up photosynthesis; meat consumption merely provides a dietary supplement. Passive traps are simple; the simplest is a horizontal leaf shaped like a shallow boat, like that of *Pinguicula*. These traps are simple in form yet complex in function. The upper epidermis of the leaf contains various types of glands, some of which secrete mucous and very sticky substances. An insect that has the bad fortune to land on such a trap is in trouble and, if it is small, might not manage to escape. It is – *mutatis mutandis* – as if it had fallen into a quagmire or quicksand, in which any clumsy, jerky movement reduces the likelihood of its escape. When the exhausted insect gets stuck on the viscous substance, other leaf glands start secreting acids and digestive enzymes, similar to those secreted by the digestive system of animals. These enzymes attack the soft parts of the insect's body, which decompose while the insect is still alive. It must be a horrible death. Other trap glands are assigned the task of absorbing the useful products of digestion, particularly compounds containing nitrogen and phosphorus. When the digestion and absorption process is complete, the insect's external skeleton remains in the trap until the rain and wind remove it.

The biochemistry of digestion is the same as that of animals: in other words, the trap is a digestive system but has no teeth, stomach, or guts

Other passive traps are in the form of cups that fill with rain water. In this environment, the plant secretes its digestive enzymes. On the rim and inner walls of the cup are hard hairs with their tips turned downwards. It is easy for the insect to go down but difficult for it to come up, and it gets trapped. These "lobster-pot" traps are very interesting as small self-contained ecosystems; their technical name is *phytotelmata*. Because these traps almost always contain food in the form of trapped insects, various other carnivorous insects that have found biochemical ways to resist the plant's digestive enzymes gather on them. The same goes for various bacteria, which also have ways to circumvent such enzymes. In this specific system, the plant undertakes the construction of the trap and contributes to the digestion of the victim along with the bacteria. Plant, bacteria, and carnivorous insects share the prey. The relative autonomy of the system allows the study of the flow of matter and energy from one organism to another.

"Lobster-pot" traps are small aqueous ecosystems called phytotelmata

Active traps operate like mousetraps

Active traps are particularly complex; they are used by the plant to catch its prey using an almost instantaneous seizing movement. The best known case is that of *Dionaea*, a small plant that is 3 to 4 cm tall and grows in the marshes of the eastern coast of North America. The trap is made up of two lobes joined at one end. In fact, the lobes are leaves that photosynthesise normally in the absence of prey. On the upper surface of the lobes are three sensory hairs arranged in a triangle. When the unsuspecting insect lands on the lobe and touches one of the hairs, nothing happens. However, if within 15 seconds the same or another hair is touched, the trap starts closing, with the two lobes rotating around the axis of their joint so the insect is sandwiched between them. In other words, the trap is insensitive to a single hit or recurring hits at significant time intervals. The adaptive value of this mechanism lies in the fact that it prevents the trap from closing unnecessarily. In essence, the mechanism has the capacity to memorize a recent event: the plant retains the information of the first arousal for a certain period and responds with a movement only if there is a second arousal within a reasonable time span. If there is no second arousal in the allotted time, the information is deleted from the plant's memory. In other words, if an insect touches only one hair and then departs, there is no point in the trap closing. Usually, however, the insect wanders around on the leaf and registers a second hit, which leads to the *coup de grace*. The trap closes and the insect panics and jerks around trying to escape, which further stimulates the hairs and accelerates closure. On the opposite side of the lobe's rotation axis are hard thorns that close the trap, like fingers when one folds his or her hands. The plot of the dramatic episode then follows a course similar to that of passive traps: Specific glands of the epidermis secrete enzymes to digest the soft parts of the insects and hydrogen cations, which make the environment acidic and facilitate digestion. The insect

Before a trap closes, it needs to perceive that a victim is present; then the consequent quick movement closes the door tightly

If the insect manages to escape in time, the trap does not close

dies a martyr's death, and the nitrogen-containing compounds resulting from its deconstruction are absorbed by other glands of the plant. When digestion is complete and only the lifeless cadaver of the insect's external skeleton remains, the trap opens again. The insect's light external skeleton is removed by wind or rain. Depending on the size of the prey, the digestive process lasts from a few days to a couple of weeks. The process is accelerated if a few drops of a carbonated drink fall into the trap, as this facilitates digestion. This system has all the physiologic features of an alimentary canal, but not its form. Sometimes, bacteria participate in the digestion process; in these instances, observation under a magnifying glass may reveal the release of tiny air bubbles.

If we facilitate the digestion with a fizzy drink, the trap opens faster

What happens if the trap is tricked into closing by several quick taps without an insect being present? It simply opens quickly, meaning that, similar to its closure, the trap's opening obeys certain signals. The signal for closing is the arousal of the hairs. The signal for opening is the completion of digestion, that is, the absence of nitrogen-containing nutrients on the trap's surface.

If the insect moves slowly, the trap closes in 1 second. If it is lively, the trap closes much faster, in about 1/10 of a second. The speed of closing depends on the number of hits the hair receives over a certain period. In any case, it is an extremely fast movement. How does it occur? What is its mechanism? In animals, the body or particular organs move with muscular contractions. The sensory organ perceives the critical information and transmits it to the central coordinating organ, where the information is compared with previous data stored in the coordinating organ's memory so it may determine the appropriate response. Is the stimulus worth a reaction? If so, the signal for movement is transmitted through neural fibres to the appropriate muscle, which undertakes the necessary action. Nothing described in this theoretic model is missing from the movement of the *Dionaea*: there is perception of the stimulus, evaluation of the information, storage in a temporary memory, and a message commanding a response, which is movement of the leaves. What is the mechanical basis for the process, though, in an organism that has no neural fibres? Let us begin at the end: the lobes move as a result of the quick shrinking of part of the cells that make up the axis connecting the two lobes. Imagine two rows of cells, one on the upper side (which will close) and one on the underside. If the cells on the upper side shrink instantaneously while the ones on the underside maintain their volume, the trap closes. How do the upper cells shrink? When the message from the hairs reaches these cells, membrane pumps are activated to release potassium from the cell. The exit of potassium also carries water out of the cell through osmosis, inevitably resulting in a reduction in cell volume. In other words, the joint of the two trap lobes has the general characteristics of an articulation. The same would happen to the human elbow if, instead of muscles, it had robust, strong cells on the external side and flexible ones, capable of reducing their volume, on the inside. If the inside cells remained turgid, the upper extremity would stay open, whereas if the cells shrank, the extremity would close. In the case of the *Dionaea,* therefore, the function of animal muscles has been replaced by osmotic phenomena.

The base of the trap has the features of a joint, and its motion mechanism is hydraulic

Still, how is the signal transmitted from the sensory hairs to the joint? For the reaction to be quick enough, the signal transmission must be equally fast. Electrical signals adopted in the animal kingdom during neural pulses also are transmitted quickly. In other words, the electrical signal progresses through successive polarization and depolarization of the membranes of sequential cells, such as those of a neural fibre. Polarization means there is a difference in the electrical potential between the two sides of the membrane. When the cells are in

The message from the sensory hairs to the joint is transmitted as an electrical signal

a resting state, there is an unequal distribution of electrical charges between the two sides of the membrane due to a correspondingly unequal distribution of potassium cations, which are positively charged. Consequently, there is an electrical potential difference of 100 to 200 μV (1 μV is one millionth of a volt), which is called the *resting potential*. A charged (polarized) membrane can be discharged (depolarized) if there is a shift of electrical charges so that the potential is reduced or becomes zero, or is reversed. The discharge of a cell causes a corresponding charge of the next one, while the first cell is recharged and returns to the resting state. The discharge then moves on to the third cell, so the electrical signal is transmitted like a wave. The same thing happens in the case of the active trap of the *Dionaea*. An electrical signal, which can be recorded with the right instruments, progresses from the sensory hair to the joint, not through morphologically clear neural fibres, but through unclear sequences of nondifferentiated cells, which also undertake other tasks besides signal transmission. We may keep in mind the basic functional similarities – as well as the differences – between the induced fast movements and the characteristics of goal pursuit in animals and plants. The similarities include stimulus perception (pressure on the hairs), electrical transmission of the signal, joint movement, biochemical digestion of the prey, and nutrient absorption. The movement, however, is not muscular; rather, it is the result of osmotic phenomena caused by the deformation of the cells of the joint, and the digestion does not require an alimentary canal.

The whole process has all the features of an intelligent reaction: detecting the signal, processing the information and evaluating its reliability, storing the data in memory, deciding whether to act, transmitting the signal, and providing movement with a clear purpose

Why does a fundamentally autotrophic organism turn to meat consumption? The explanation lies in the habitat of carnivorous plants. They grow in marshy areas or, generally, in soils in which nitrogen is scarce. When the soil floods, the water sends off oxygen through soil gaps, and soil conditions become anaerobic; this favours the growth of anaerobic bacteria and

For plants, being carnivorous is a way to supplement their diet

inhibits the aerobic ones. Many anaerobic organisms use a kind of respiratory metabolism that turns all nitrogen-containing compounds in the soil into molecular nitrogen escaping into the atmosphere. Apparently, the scarcity of nitrogen in the soil has exercised an evolutionary pressure favouring the selection of carnivorous habits among some plants that are photosynthesising in every other aspect. Meat consumption, therefore, ensures a dietary supplement that improves the adaptability of these plants in this specific habitat.

Carnivorousness is a rare phenomenon in plants: only one in a thousand known species has adopted meat consumption. In theory, this is not impossible. Many plants have the information (i.e., the gene packets) needed to produce mucous polysaccharides and to construct gland cells that secrete substances on the surface of the plants' bodies. Nectaries secreting highly concentrated sugar solutions and adenoid hairs that secrete essential oils are two such examples. Frequently, small insects become trapped on mucous plant surfaces and lose their lives there, without the plants being carnivorous. Because the dead insect inevitably will be decomposed by bacterial action, such plants often are characterised as potentially carnivorous. The products of decomposition may be absorbed by the plant and meet part of its nitrogen requirement. Plants with viscous surfaces also may be considered precarnivores, which with some more mutations or gene thefts might become real carnivores in the future. It also is possible that the number of carnivorous plants may be larger, because no one knows exactly what is happening underground, in the half of the biosphere that still holds many secrets. Recently, the case of a small plant named *Genlisea* was reported; this plant is rootless but has small subterranean modified leaves with holes 200 to 300 μm (0.2–0.3 mm) in diameter. Tiny protozoans and crustaceans (small shrimps) are chemotactically attracted to these holes by substances secreted by

Many plants have the right gene packets to be "potentially" carnivorous

Subterranean meat consumption by plants

the plant. Entering the hole is easy because the hairs around the opening bend inwards but not outwards; therefore, the animal becomes trapped and is digested by enzymes secreted by the plant.

Plant Movements

The case of carnivorous plants is not the only one that violates, to a certain extent, the basic principles of plant biology. The term *to a certain extent* is used because carnivorous plants use this method to supplement their diet and do not give up photosynthesis. However, in the case of active traps, it appears the principle of immobility also is violated, again to a certain extent, because these plants do not hunt their prey or move in space but make fast movements in a part of their bodies. They merely are equipped with a fast-moving snare. Are there any other cases? What purpose might they serve?

When the principle of immobility is violated

Mimosa pudica (also known as *touch-me-not*) is well known for moving its leaves in response to being touched. The characteristics of this movement are similar to those of the *Dionaea* trap: electrical signal transmission from the point of contact at the base of the leaf, activation of potassium pumps in certain cells of the "joint," water loss, and osmotic deformation of the cells, which instantaneously moves the whole leaf downwards. Although the mechanism is the same, the adaptive value of the movement is different. The movement frightens the insects that land to feed on the leaves of this small plant. The plant does exactly what humans do when they feel an annoying fly touching them. Just like a human, the plant feels the pressure on its epidermis and acts accordingly.

Mimosa pudica whisks away flies, as horses do

Quick plant movements are impressive and easily perceived; they are explained by a cause–effect rationale, that is, they are provoked by an external factor and

Movements with no apparent external stimulus: the case of climbing plant tendrils

have a specific result. However, other plant movements are not as fast and occur independently, with no apparent external stimulus. These usually are growth movements – the plant's organ moving also increases its size. An example is the movement of climbing plant tendrils. They may appear immotile, but they move continuously and not at all randomly. They seek a target, a fixed object around which to tightly wind and support the rest of the plant. How does this happen?

Tendrils move like a screw into the air…

Tendrils are long, modified leaves or shoots with a round cross-section. Filming them and playing the film in slow motion reveals that their ends move in a circular fashion. However, the tendril keeps growing in length, which means the propelled end moves in a spiral, as if being screwed into the air. How is this motion achieved? Imagine the cells of the upper side of the tendril growing faster than those of its underside. This would result in its turning downwards until forming a ring, but this is prevented because the zone of growing cells keeps shifting by circling around the tendril's longitudinal axis. The growth zone shifts clockwise – like a screw – seeking a fixed supporting object. When this object is finally found, the cell growth zone stops shifting and locks in the side of the tendril that is not in contact with the supporting object. In this way, growth is limited to the side opposite the touching surface, and the tendril rotates around the object and consolidates the contact. Limiting the cell growth zone not only concerns the part of the tendril in contact with the object, but also extends to its base. This results in a differential growth rate on the tendril's external side. A spiral is formed, which, like a spring, pulls the shoot of the climbing plant towards the supporting object. When the process is complete, the walls of the tendril cells are lignified so as to become strong and resistant.

…until they reach a secure supporting object

The circular "screwing" movement turns into a guided one only after the tendril has "hit" the supporting object repeatedly. If the support is artificially moved, the motion continues to be circular. The tendril cannot be cheated; it demands some guarantee that the support is permanent, and does not change its behaviour until it receives such a guarantee. As in the case of the active trap of the carnivorous plant *Dionaea,* the tendril stores the information of the initial contact in a temporary memory bank and does not change the direction of its movement until it perceives additional touches on the same side. Obviously, the epidermal cells of the tendril perceive the hit as though they had a sense of touch. The tactile signal is processed, and when its reliability is confirmed, the message for changing behaviour is passed. Although the mechanism through which the growth zone shift stops is not known, it might be related to the awakening of genes that selectively inhibit growth on all sides except the one diametrically opposed to the supporting object. Furthermore, genes related to lignin synthesis also are awakened so that the tendril may become lignified. Similar to so many other processes described, this one also indicates an intelligent reaction with adaptive value on the part of the plant. In this manner, plants with tendrils counterbalance their lack of a strong stem. They are propelled towards the light at the expense of other plants; they face wind pressure using others' forces and do not collapse under their own weight. In the broadest sense of the term, they may be characterised as parasites.

Successful tendril anchoring requires the memorization of suitable tactile stimuli. This is another intelligent reaction.

Climbing plants, just like many humans, enjoy provisions supported by others' skills

Chapter 10
Are Plants Intelligent Organisms After All?

Towards a Biological Definition of Intelligence

Anthropocentric and biological definitions of intelligence

Are plants intelligent organisms? The question would be a real challenge if this chapter were placed at the beginning of the book. However, we now have a basis for discussion if we can also manage to define *intelligence*. Although many books have been written on the subject, experts have not yet reached a general consensus on the definition. There also is much debate concerning attempts to quantify human intelligence with what are known as IQ (intelligence quotient) tests. Most dictionaries define *intelligence* based on psychological models, while remaining limited to human measures, and stress qualities such as the capacity for reasoning and abstract thought or even aesthetics. Moreover, they emphasize that the aforementioned traits are linked to cerebral functions, further implying that intelligence is an exclusively human feature. Still, for our discussion to proceed, we need a definition based on biology, given that the question can be raised in regard not only to plants but also to other organisms. Do dogs, worms, and frogs have intelligent reactions?

Intelligence as the possibility of sound perception of the actual world

Some dictionaries, liberated from an anthropocentric attitude, define *intelligence* as the potential for sound perception of the actual world. This exceptionally succinct definition is much closer to the initial etymologic interpretation of the term, in both Greek and Latin. *Euphyis (*Ευφυής*)*, from *eu-* (εύ- meaning "well") and *fyomai* (*φύομαι,* meaning "to grow") is an organism that grows well. Because Greeks were aware of the unity of body and mind, *euphyis* came to mean "astute," "clever," "capable of successfully adapting to the environment." The Latin term *inter legere*, from which the English *intelligence* derives, originally meant "the act of gathering from between," that is, "to pick out or discern."

Physiologic stages of perceiving the world among organisms that have a central coordinating organ

How do we soundly perceive the world? One way is through the senses, but this is hardly enough. Every piece of information from the environment that reaches and stimulates a sensory organ is transmitted to the central coordinating organ (the brain), where the significance of the information is discerned through comparison with other, similar data recorded in the memory, that is, experience. Up to that point, knowledge is acquired – the brain becomes conscious of meaning and assesses information. At the same time, however, a decision is reached as to whether the organism needs to respond to the stimulus. As I write this, I can hear the wind blowing in the trees, but my judgment says I need not react: the information collected through the sense of hearing is recognised by the brain as wind and causes no alarm. If, however, the phone rings, I have to react; the command from the central coordinating organ is "get yourself to the side table *now* – the message might be important," whereas on my way to the phone, the coordinating organ is trying to predict who might be calling at this time of day and what he or she might want. The manner of responding to the stimulus – what happens after the message is processed (i.e., it is comprehended) – is at the core of intelligent reaction. Assume that the photoreceptors of an antelope's eye receive the signal of a brown four-legged animal with a mane and sharp canines, while its eardrums are stimulated by the wave frequency of a beastly growl. If the antelope does not comprehend that this is a lion, something is wrong with its perception. If the perception is correct, in other words, that it is a lion (and indeed one getting threateningly close) and the antelope does not start running, something is wrong with its intelligence. It will not survive, and its genes will not be part of future generations. At the same time, however, the antelope has to weigh the scale of the threat.

Response to stimuli is intelligent when it strengthens the organism's adaptability and fitness

If the object – definitely a lion, judging from its sight and sound – has already gorged itself on other prey and decides not to spend any time with the antelope, there is no point in the latter running. It would become the laughingstock of the herd, and the waste of energy would leave it unnecessarily tired and unable to face another predator – which might be hungry this time – further down the road. Therefore, sound perception of the actual world is not enough; there also must be an appropriate decision that is of adaptive value – by increasing the individual's likelihood of survival and therefore the probability of producing offspring and perpetuating the species.

Plants as intelligent organisms, in the biological sense of the term

Are there similar phenomena among plants, that is to say, phenomena that entail collecting environmental stimuli, processing and assessing information, and reaching a decision to change one's behaviour? If so, what is a plant's "behaviour"? If the behaviour of a plant does indeed change, is this change of adaptive value? In other words, does it increase the individual's likelihood of surviving and perpetuating its species? Also, because the processing of information presupposes memory, is there memory in plants? Finally, can there be intelligent behaviour when there is no brain involved? There absolutely can. Let us recall briefly some of the plant biology phenomena already described, which now will be examined based on the criteria for probable intelligent behaviour.

Although this discussion is limited to plants, the reader should note that all organisms, however simple and passive they may appear at first glance, display intelligent behaviour as defined earlier. If not, they could not survive.

Examples of Intelligent Plant Reactions

Not only is there intelligence without a brain, but, as mentioned in Chapter 6, there is vision without eyes. Plants perceive radiation of various spectral regions, with at least two to three photoreceptors, which – from a functional point of view – function like rhodopsin in human eyes; the only difference is that instead of being located specifically in an "eye," these receptors are scattered throughout almost all the plant's cells. As has been demonstrated, all plant functions are decentralised and their coordination is not an exception to this rule. Every cell has an independent system of sensory organs. One of the plant's photoreceptors – the phytochrome – is stimulated, that is, it "sees" light in the red (630–680 nm) and far red (700–750 nm) spectral regions. The far red region is hardly visible to the human eye. The spectral region from 630 to 750 nm is very important for plants, because it contains wavelengths intensely absorbed by chlorophyll (red) and those hardly absorbed at all (far red). Proper signal processing (i.e., recognising the ratio of red to far red photons) allows the plant to determine whether it has neighbours and, if so, whether they are above or below, next to it on the right or the left, or taller or shorter than itself. Every neighbour is a prospective competitor for environmental resources, such as light, water, and soil nutrients. This means that such information is of immediate, vital importance. What the plant continuously monitors is whether there are objects in its surroundings that absorb red light with their chlorophyll without absorbing (therefore, reflecting) far red light. The critical issue here is whether the plant's locating a neighbour leads to a decision to change its behaviour and whether this change is of adaptive value, that is, whether it improves the plant's quality of life and its survival odds.

Decentralised "vision" without "eyes"

When is an optical piece of information vital for a plant?

What constitutes behaviour in the case of plants? In the animal world, behavior changes usually are perceived as movement, as in the antelope example given earlier. The speed of animal movement is such that it falls within our range of perception – or, to be more accurate, from an evolutionary point of view, natural selection has determined the range of perception of movement within the limits of the actual movement potential of competitive organisms – or organisms interesting for other reasons – under the specific conditions of planet Earth: neither too fast nor too slow. We perceive that an object revolves if the revolution frequency – the time it takes for a discernible point of an object to pass in front of our eyes twice – is longer than the time that elapses between the visual signal of the discernible point captured by the eye and the point at which the brain comprehends what it is about and registers it as information. Objects that revolve much faster appear immobile. The same is true for objects that move too slowly. They do not attract attention because they are not dangerous. This is the case with plant behaviour. Often, the intelligent modification of a plant's behaviour in response to an environmental signal is so slow that it is not perceived as such. One such example is the plant's response to identifying a potentially antagonistic neighbour by turning towards the opposite direction and/or gaining height in search of more light. Furthermore, this leads to internal redistribution of resources, which are now directed towards growing upwards rather than in other directions. We are aware that both movements (i.e., upwards and away from the competitor), which are too slow for the human eye to perceive, have adaptive value because they secure better light exploitation for an immotile organism fixed to the ground, for which light is an energy resource. According to the definitions, these are intelligent responses.

Natural selection has not given humans the ability to perceive very slow movements, probably because they do not constitute a threat for them

Plants make very slow growth movements when they sense they have neighbours

Two cases are worth examining in detail. The first case is that of an already established plant that has no neighbours. It receives natural white light from the sun and reflected light from the ground, in the spectral frequency distribution characteristic of the soil. Inevitably, other plants appear and grow below it. The light reflected from below is now enriched with far red and depleted of red frequencies. In other words, the light has changed colour. This change is perceived by the first plant, and an alarm signal is initiated. If neighbouring plants grow faster than the first plant, they soon will cast their shadow over it. The choice is obvious: it must start growing faster, modify its form, and become even taller. Corresponding internal hormonal changes direct the use of resources towards the new target action. The first plant continuously monitors the growth rate of its neighbours so as to correct both the direction and pace of its own growth accordingly. The capacity to grow asymmetrically is an integral part of plant development under these conditions. In this specific case, it is important not only to monitor the position of greenery in the surroundings. The plant, in essence, assesses the current situation while anticipating the future and modifies its form accordingly. An animal's behaviour mainly entails fast movement; a plant's behaviour is to change its form. This phenomenon, described in Chapter 2, is called *phenotypic plasticity*.

In essence, plants measure the quality of light reflected off neighbouring plants and change their own behaviour

Anticipating the future by continuously monitoring the present

The second case is that of a plant in the undergrowth of a forest shaded by other plants. The light it receives penetrates the cover of overlying plants through the gaps of less dense branches and foliage. This light environment is exceptionally complex because gaps are not evenly distributed and the light keeps changing with the daily course of the sun. In other words, there is a statistical distribution of light – more at some points and less at others – that changes in ways determined by the structure of the cluster. Our plant has to estimate where to grow its branches and how to turn its leaves.

The light environment in the forest understorey is spatially and temporally complex

It has to map the light environment and model its response; apparently, it manages successfully. All relevant studies indicate that its growth pattern is not random. The light environment, however, is not stable in the long run. A tree shading our plant might get sick or fall victim to thunder or human intervention. The gap created changes optical data and requires revision of the initial growth pattern. The same is true if the forest is deciduous and seasonally loses its leaves, clearing the sky above. It now can be better understood why plants do not have a stable shape and size and why their form is characterised by such plasticity; why no plant of the same species is the same as the one next to it; and why it is necessary for embryonic tissue (the meristem) to always be present in plants to serve their frequent need to grow new branches.

Mapping the light environment and its changes, and the consequent "decision" to place branches and leaves accordingly

In the previous examples, the stimulus for behavioural modification is a change in the quality of the light. In other words, the decision is based on assessing a single environmental parameter. The system is simple, considering that, generally speaking, animal organisms make decisions after they process several different bits of information. The antelope combines visual, sound, and olfactory signals, all arriving at the central coordinating organ, which decides whether there is good reason to start running. Are there decisions for behavioural changes in plants based on the assessment of more than one factor? Can plants make combinations?

Recall the example of the stomata described in Chapter 6. Stomata are small valves, invisible to the naked eye, on the epidermis of leaves. Through them, gas exchange occurs between the leaf and the atmosphere; what enters is mainly carbon dioxide (CO_2) to participate in photosynthesis. Inevitably, it is through these same valves that water (H_2O) vapours escape to the external environment. The opening of the valves, therefore, is of vital importance. A large opening means

When numerous factors are co-assessed

faster inflow of CO_2, more intense photosynthesis, and faster growth. However, it also means greater water loss, which, if water is not abundant, must be resolved before it leads to dehydration. This is a major and complex dilemma, because CO_2 photosynthesis into sugars fluctuates depending on light intensity; it is potentially higher around noon on a cloudless day, lower in early morning and late afternoon when there is less light (as well as on a cloudy day), and interrupted at night. However, around noon on a cloudless day, the greatest water loss also is observed, because the natural phenomenon of evaporation from the leaf interior to the atmosphere follows the laws of physics. Not only does it increase with temperature, but also with lower relative atmospheric humidity, which also coincides with the periods in which daily temperatures are at their highest. Observing necessity, stomata close at night. When there is no photosynthesis, there is no point in risking the loss of precious water. However, during the day, the opening of the stomata is not stable, but changes depending on the conditions. The rule is for valves to open to such an extent that the ratio of CO_2 entrance (photosynthesis) to water loss (transpiration) is optimal, that is, there is sufficient photosynthesis with minimal water waste. How do plants resolve such a complex issue?

Stomata: leaf valves with pressing dilemmas...

...which, however, resolve complex issues

The cells comprising the valves (called *guard cells*) have three different types of sensors. The first type provides information about light intensity. If there were only this one, the stomata would be closed at night, but during the day their opening would follow the daily fluctuations of light intensity. The second type of sensor monitors the concentration of CO_2 in the leaf interior; in essence, this provides information about how fast chloroplasts remove CO_2 from the leaf interior during photosynthesis. The initial information "there is light" is complemented not only by information that the light is being used but also by information on how effectively it is used. For example, a temporary or

Stomata continuously measure three environmental parameters and an internal hormone

more permanent inability of chloroplasts to photosynthesise (i.e., to absorb CO_2) satisfactorily despite the presence of light increases CO_2 in the leaf interior and sends a message to the valve to limit its opening so water is not wasted in vain. The third type of sensor monitors the relative humidity of the atmosphere. If it is high, the valves can open more, because water evaporation is slow in a humid atmosphere; in contrast, evaporation is intense in a dry atmosphere. Therefore, the two sensors (for light and humidity) transmit to the valves from the external environment information regarding the favourability for photosynthesis and unfavourability for water loss. The third sensor (for CO_2) monitors the current efficacy of the photosynthetic apparatus, that is, the internal conditions.

The information perceived concerns light intensity, the efficiency of the photosynthetic mechanism at any given time, and humidity in the atmosphere and soil

The three aspects of information are assessed at the guard cells, and a decision is made regarding the valve-opening size so the ratio of photosynthesis to water loss does not deviate much from the optimal point. It is as if the guard cells have to solve an equation that determines the final valve opening on the basis of three independent variables. The result is that the opening of the valves is never stable but keeps changing throughout the day and that, as seasons change, decisions may be modified at any moment. However, the foregoing is true to the extent that the water evaporating into the atmosphere is satisfactorily replaced by that provided through the root. What happens, though, when the root cannot replace losses, that is, when atmospheric conditions send a message for the valves to open, but there is no water in the soil? In this case, the root perceives (through its own sensors) that the soil water is not sufficient; it produces and sends through the vessel system a chemical message in the form of a hormone to the valves; this message overrules all previously sent messages. It operates like a switch for the mechanisms opening the valve, which makes the stomata close.

Therefore, the problem for a plant with sufficient water is the size of the valve, that is, the assessment of photosynthetic gain compared with dehydration risk. When the water is insufficient, though, the dilemma becomes "photosynthesis or survival." Should the plant grow or survive? Elementary intelligence points to the second option, which is imposed by the root until better days come and the alert is over.

Information is processed within stoma cells, and a "decision" is made as to what its optimal opening size should be

Indisputably, the regulation of valve opening has all the characteristics of a biologically intelligent reaction: many stimuli are perceived, there is a flow of information that is processed and assessed, a decision is made, and the behaviour is modified successfully. All these events occur in the guard cells, which make up the valve. In the case of the stomata, however, control of gas exchange is not just short term, for the day or season. At a different level, it is also a long-term need. Photosynthesis and transpiration depend not only on valve opening but also on the number of valves. The number of stomata – or their density – is determined early in the life of a leaf, when it is young. At selected points of its epidermis, epidermal cells are differentiated into guard cells. It has been found that the density of stomata is inversely proportional to the concentration of CO_2 in the atmosphere. Less atmospheric CO_2 leads to a larger number of stomata. The adaptive value is apparent. When CO_2 in the atmosphere is abundant, photosynthesis is facilitated; therefore, there is no need for numerous valves through which water would be inevitably lost. Because the density of stomata is determined in the early stages of a leaf's life, it may well be claimed that the plant – *mutatis mutandis*, as always – makes assessments on the basis of CO_2 levels at that time as to how many stomata it will need in the future, when its leaves are mature, so that photosynthesis is sufficient but water loss is not excessive. In terms of biology, that is, on the basis of its lifestyle and *modus operandi*, the plant

An a priori estimate as to the number of stomata future leaves will need

anticipates what its future needs might be. What is expected, therefore, is that based on the imminent climatic changes and the increase of atmospheric CO_2, future plants might resolve the lack of water more efficiently.

What happens to a branch in deep and permanent shade for a long period? This rarely happens in nature and can easily become the subject of an experiment. In economic terms, this means the branch is not productive, does not photosynthesise enough, and is a liability for the plant because it consumes (through respiration and maintenance needs) more than it produces. If this unpleasant situation continues for a long time, the plant kills the branch through food deprivation: communication bridges are cut by sealing off the vessels that transport water, mineral nutrients, and carbohydrates. The useless branch falls off the tree and is replaced by another one at a more favourable and better-lit location. Although the details of this mechanism are unknown, it must be assumed that it could not happen without detailed spatial perception, accurate data processing, and successful decision making, which increase the organism's adaptability.

Any plant part that becomes nonproductive is rejected without any second thoughts

The expression *decision making* should not be considered odd. Woe betide the organism – whatever its level and scale of classification, from the humblest bacteria to humans – that cannot perceive its environment, respond to challenges, and make decisions. If intelligence, in essence, means the capacity to adapt by changing one's behaviour, then it needs to be exercised at any given moment, taking into account the lifestyle of each organism; this is the measure of its success. Although plants are not capable of abstract thought and do not read or listen to music, they have the intelligence they need to survive, reproduce, and expand, as well as to make decisions following careful analysis of the situation.

Cuscuta is a parasitic plant; it has no leaves or chlorophyll and does not photosynthesise. However, it roots, flowers, and reproduces like any other plant. *Cuscuta* is not an autotroph but lives as a parasite at the expense of other plants. It wraps itself around other plant stems and sucks nutritious organic substances from their phloem vessels. If one is patient enough to observe the fine, rosy-yellow shoots of *Cuscuta* when it is searching for the right victim, he or she will see that in many cases, the shoot of the parasite unfolds from the stem of the victim some hours after their initial contact. It is as if the parasite abandons the victim before enjoying its juices. Is there a preference for specific shoots of the victim? If so, what is the basis for it? This is not a random phenomenon; the degree of rejection is lower in potential victims containing more nitrogen, an important chemical element required by all organisms. What happens is that the parasite shoot investigates the quality of the victim before growing suckers, that is, before investing in a permanent establishment. Once it has established itself, the contact surface area between parasite and victim is larger in proportion to the victim's nitrogen content. Similar reactions among people are characterized as intelligent without the slightest hesitation. No sensible person buys anything before making sure the product is of proper quality, and nobody makes an investment without weighing the long-term benefits. When one lacks basic productive capacities (such as *Cuscuta* in the world of plants) and decides to become a "parasite," it would be stupid to select a poor, sick host instead of a rich, healthy one. Such behaviour is considered intelligent among people, so why should we not characterize it as such in the case of plants? *Cuscuta* samples the quality of the products and chooses the best. Based on this qualitative assessment, it decides on the extent of its investment. In its own (still unknown to us) manner, the parasite calculates the cost and potential benefits.

How the victim of the parasitic plant Cuscuta is chosen

In essence, Cuscuta assesses the quality of the potential victims, rejects the inferior ones, and invests in "suckers" on the plant presenting the best option

When humans act in a similar manner, they are characterised as intelligent. Why should this not be the case for plants?

Within the framework of its own lifestyle, it makes a successful entrepreneurial investment.

Recognition and Decision Making

Why does Cuscuta not become a parasite on all plants? Why not on itself?

How do plants recognise themselves? Are they aware of their own individuality?

Cuscuta does not become a parasite on all plant species, which means it has the capacity to recognize the right hosts and possibly those that are less resistant. Furthermore, it does not become a parasite on itself, meaning it recognizes itself. Is it too much to say it is aware of its own individuality? Usually, there is a spontaneous, reflexive negative response to such questions. All the relevant literature on humans considers only that species to be self-conscious; this probably is true if viewed in the context of a species that has developed literary works, abstract thought, philosophy, and metaphysics, a being that is aware of its finite existence. Still, the lesson from the previous discussion is that such questions should account for the lifestyle and *modus vivendi* of each organism, as well as the nature of its vital interests. Therefore, are plants aware of their own individuality and in what way? Do they recognize themselves?

Important information on this issue is provided by studies of the hidden and – until recently – unknown part of plant behaviour, that is, the behaviour of the root system. The underground world is particularly competitive. Roots fight to occupy vital space, allowing them to better exploit soil resources, that is, water and the necessary minerals. As already mentioned, the development of the root system is mainly centripetal (directed towards the centre of the earth) but with a significant collateral parameter. The density of the roots depends greatly on the availability (abundance) of resources. The root perceives regions locally rich in resources through continuous chemical investigation of its environment. Roots follow the gradient of nutrients to their source, and new rootlet growth is induced on the spot to increase the absorption surface. In other words, root density is high in the vicinity of nutrient sources. This happens in the absence of competition. What happens, though, if the root encounters that of a rival? It changes direction. The signal is not tactile; the directional change occurs before physical contact is made, obviously based on chemical signals. The method is more economical if the plant does not wait until it makes actual contact with its competitor and then avoids it; it becomes aware of its rival by perceiving the chemistry of its secretions, a process similar to olfaction. Therefore, plant roots are denser where there are no other plants and become sparser in the presence of other plant roots. This means the plant perceives its neighbours and responds but does not respond similarly to its own secretions, to which it is immune. In this sense, it recognizes what does and does not belong to it, which products are from its own genes and which are from another's.

Underground competition

Plants recognise their own and other plants' roots

Recall a similar case described in the discussion of plant reproduction. The wind or insects deposit on the flower stigma a multitude of pollen grains from a range of plant species: from other species, from other individuals of the same species, and from the same individual. Fertilisation cannot take place with pollen from individuals of other species and should not take place with pollen from the same individual, as self-pollination is to be avoided. Only pollen grains from another individual of the same species germinate and enter the stylus. How is the origin of the pollen recognized? Gene products, mainly proteins, are detected. When there is no correspondence whatsoever (different species), germination is inhibited. If there is total identification (same individual), germination also is inhibited. When there is the right correspondence, germination is allowed. Sometimes, fertilisation is not

Weighing potential lovers on the basis of their gene products

even allowed between close relatives. In poppies, for example, it has been observed that fertilisation by the pollen of neighbouring plants is less frequent than that by the pollen from plants at some distance. Neighbouring plants are more likely to be relatives, as the dispersal of poppy seeds does not cover a wide range. The test, therefore, is very precise, albeit less intense at times. When a plant finds itself in reproductive isolation and there are no other individuals of the same species around, then instead of remaining nonfertilised, it turns a blind eye and accepts its own pollen. However, not all plants do this. Most, as is well known, can spend a long time reproducing vegetatively (asexually). Others, when in a tight spot, reduce their demands, bypass the test, and accept pollen from other species, provided they are related. This is how hybrids are created, which are fertile in plants, as opposed to animal hybrids.

However, when good lovers are scarce, the second best benefit

In both cases of self-recognition (root competition and fertilisation), what is recognized is the exact chemical nature of the products of the individual. This is similar to what our immune system does. It recognizes the chemical quality of our cells and that of alien cells, and attacks the latter. Note that in the case of plant fertilisation, although the chemical test is the same, it may lead to different behaviours, depending on the circumstances. If there is ample pollen from different individuals of the same species, the plant excludes its own pollen as well as pollen from plants of other species. If pollen is scarce, the plant makes do with its own or that of other species, provided they are not total strangers. This is not a full collapse of morals (i.e., a total collapse of chemical prohibition) but a selective mitigation. When the population returns to a normal density for the particular species, traditional sexual life is restored.

In other words, for the same test result, behaviour might differ, allowing for a selective change in attitude

It is not known how prohibitions weaken or intensify. Reasoning leads to the conclusion that the system probably also measures other things and is flexible enough to allow deviations from and exceptions to the rule. I use mild words to reach the point I eventually want to make: that, in essence, this is decision making. A signal is received, information is weighed and evaluated, and a decision is made to change behaviour. This change benefits the plant and its longevity, even in the form of a hybrid or hermaphrodite. After all, the topic of this discussion is self-recognition.

Is this decision making?

Let us return to the subject of a root's recognizing alien roots and avoiding them by changing direction. One may claim this is an instinctive, rigid behaviour pattern that instructs, "Proceed normally and turn only when you encounter an environment chemically different from yours." In other words, the root operates spontaneously, in a predetermined manner, without deviations. There is no question of "decision" in such a case. One easily may refute the view that things are this simple based on the information presented in Chapter 6: for example, a root changes direction if it encounters a hard object, such as a stone. In this case, it does not recognize a chemical but a tactile signal in the form of pressure. It does not persist in vain, trying to penetrate the obstacle, although many plants do so if they find no other way. A stone usually is circumvented. Quite often, the root has to deal with contradicting signals, for example, repelling signals from other plants and attracting signals from regions with a high density of nutrients. Its response, therefore, is not an either/or reaction, which would have been absolutely predictable, but the result of contemplative actions assessing multiple signals. Even more important is the fact that the reaction to avoid other plant roots is not always the same; it depends on the soil volume available. Experiments in limited spaces (e.g., flower pots) showed that, in this case, the tendency for avoidance turns into the complete opposite: plant roots are directed towards each other in an obvious effort to exploit the

The root perceives chemical and tactile signals, which often are contradictory

The root measures available space…

...and, accordingly, might exhibit conciliatory or aggressive behaviour

largest share possible of the limited soil reserves, of which the rival will be deprived. This indeed is a 180° turn, a radical change in tactics. Because avoidance of alien roots has no adaptive value in the specific case of the limited volume of the flower pot, direct attack becomes the preferred *modus operandi*, and the one that can exploit it will gain control over the whole pot.

If someone gets in your way and there is ample space, you likely would prefer to move to a more comfortable spot than to start a fight that will spoil your mood and interfere with other, more productive activities. However, if you realize that this annoying individual is threatening your vital space, you would weigh the pros and cons of a battle or a tactical retreat. If the adversary is intransigent and adds fuel to the fire, the rupture is inevitable. In this anthropocentric description, the concepts used include conciliatory or aggressive behaviour, intentional change of behaviour, and restoration or disruption of living in concord. Why are we hesitant to use the same terms in the case of plants? Why not describe root behaviour using the same concepts?

Memory?

Let us conclude this chapter on plant intelligence by raising the issue of memory. Do plants exhibit some sort of behaviour that indicates memory? Following are four noteworthy cases, two of which have already been presented. In climbing plants, the twisting, growing tendrils follow a spiral course as if screwing upwards into the air. If they encounter a stable object, they do not fold around it immediately. They require a second or even a third collision on this stable point before the decision is made for the tendril to wrap itself permanently around the support detected. Tendrils are not cheated into wrapping themselves in vain around an object that is not stable. The impact signal is tactile: they perceive pressure. The fact that they need the next signal, from the same stable position of the support, means they maintain the information of the initial contact in storage for some time. If there is no second signal within a reasonable time interval, the first stimulus is deleted from the temporary memory. If the signal is repeated soon, there is some warranty of stability from the support and the tendril's behaviour changes. Similarly, insect-eating plants of the *Dionaea* species do not close their traps until they receive repeated tactile signals on three sensitive hairs within an interval shorter than 15 seconds between each successive contact. This ensures that the trap does not close in vain when the insect that landed leaves immediately. The duration of temporary memory in this case is around 15 seconds. The manner of information storage is not known.

Tendrils of climbing plants

Active traps in carnivorous plants

There are also two cases that may be considered long-term and, in some way, permanent memory. When some plants grow for a long time under conditions of mild water scarcity, with water provided abundantly once a year, they attune to their growth period (e.g., the production of new leaves) with the watering season. Indeed, they initiate their growth a little earlier than the actual watering, as if they remember or anticipate the favourable period soon to arrive. Another case studied concerns trees in the north, which every few years receive a seasonal and intense attack of leaf-eating insects that completely denude them of foliage. The year after one such attack, it was observed that, at the same time of year they had suffered the previous attack, the trees increased the levels of insect-repelling phenolic compounds in their leaves, even though there was no actual attack. It was as though they remembered what they had suffered and were preparing for the likelihood of another invasion.

Short-term and long-term memory

Dissimilar things cannot be compared with each other

Plants have the intelligence that suits their own needs

An IQ test made to measure the intelligence of Amazon native tribes would prove an average university professor a moron

The reader may still be suspicious and think the author is exaggerating. He or she might be justified in doubting whether the plant reactions described can be compared, from a qualitative point of view, with fine human intelligence. However, this book does not make that claim; it would be naïve indeed. No plant (just as no animal) has ever comprehended human nature. Humans have the intelligence to understand the nature of plants. It would be a shame not to use this intelligence in that direction. Therefore, it is hoped we can agree that each organism has the intelligence suitable for it, an intelligence that characterizes its *modus vivendi*, lifestyle, and strategies and that helps it face its own problems, find its own typical solutions in the context of its biological position, and maintain its livelihood successfully. Human intelligence-measuring systems are not applicable even within our own species. A high IQ, which might lead to a successful career in a Western urban environment, would hardly help in the Amazon tropical rainforest, the Australian desert, or the arctic region. If native residents of these regions were to invent their own system of measuring intelligence and their own questionnaire, suitable to their environment and needs, our performance on that test would classify us as morons.

Epilogue

A Tribute to Darwin

The author is happy if he has managed to help readers see plants in a different light, if he has managed to convince them that plants are organisms no less complex than animals or humans, or, rather, that they are as complex as their *modus vivendi*, which is totally different from ours, requires. A plant's lifestyle is characterised by slow movements of the parts of otherwise immotile organisms, which produce its own food using simple raw materials abundant in nature. Plants are organisms whose waste products not only do not constitute a problem, but actually contribute to creating soil and maintaining the atmosphere in its current oxidising state. The state of the atmosphere – with enough oxygen to allow animals efficient energy exploitation of their food, which therefore facilitates their movement and growth to a satisfactory size, yet not enough to oxidise these organisms – is a by-product of the plant activity called *photosynthesis*. Photosynthesis is a unique, wide-scale *modus operandi* of the biosphere to exploit a sustainable extraterrestrial energy source – the sun – and transform light energy into chemical energy. Plants then use this energy to produce complex organic food from simple raw materials abundant in their environment; it is cheap food used by all and capable of sustaining the whole ecosystem. One might put forward that humans, through photovoltaic panels, have managed to exploit solar energy by transforming it into electricity, but that would be a hasty response. Photovoltaics cannot reproduce, maintain or repair themselves, or propagate on their own, and when they break down, there is no way to dispose of them because they do not decompose to become soil and fertilizer as leaves do, nor have any microorganisms been found yet to undertake this task free of charge.

The author also hopes it is now clear that it is degrading for plants when one refers to a person who is not in touch with his or her environment as being in a "vegetative state" or a person not famous for his or her intelligence as a "potted plant." By simply scanning the marginal notes in each chapter, readers may refresh their memory and find arguments against such prejudice.

Indeed, the marginal notes may well be a substitute for this epilogue. However, the author feels there is still something missing, an issue pending that must be discussed, a point readers might wish to have addressed. How have we reached the point of having all this knowledge about plants today? What type of activity, what motivation, and what human qualities have made such knowledge possible? Who are the experts who generate relevant knowledge, and how do they work?

I do not wish, nor is it my intention, to present biographies. In any case, I do not think that today an independent observer could discern – at either the university or research institute level – an archetypal scientist-researcher with a set of intellectual and behavioural traits prerequisite for entering this field of research honorably. Today, it is a wide-open field, a vocational realm like all others, in which one may find all kinds of individuals: useful and useless, consistent and frivolous, intelligent and less intelligent, enthusiastic and apathetic, industrious and indolent, honest and dishonest, modest and vain, those made of gold and those made of tin. The results of this lack of criteria for entering the field and the creation of a crowded, regimented, social group of professional scientists, with a specific legal form and trade union demands, will be studied by history, which will accurately weigh the contribution of this group of people against the scientific progress and prosperity of humankind. However reluctant, the author cannot seem to reject the thought that the more the number of people entering this field increases, the less likely a major discovery becomes. Thousands of scientist-researchers, particularly in recent decades, although starting with the best of intentions and probably the noblest of ambitions, are suffocated by routine practices and end their careers with a poor and disappointing "statement of accounts." What is it that makes some individuals differ and leave behind – *mutatis mutandis* – a trace for all time?

Readers reasonably, and not unjustifiably, may assume that those characterised by the first adjective in the long series of opposites presented a few lines earlier are more likely to enjoy, with a little luck, some form of scientific immortality. The same is true about all vocations. The author must admit that these thoughts are not random: they emerged from wondering about such questions and because of specific circumstances. It so happens that in 2009, the scientific community celebrated a double anniversary: the 200th birthday of Charles Darwin and the 150th anniversary of the first publication of the most debated scientific theory in the history of biology, the one Darwin expressed in his book about the origin of species (by means of natural selection). The significance of this book for modern biology will not be analysed here – that analysis has been done by others far more competent. However, a few brief questions may be raised concerning some of the character and personality traits of a scientist who certainly will be celebrated on similar occasions a thousand years from now. What kind of scientist was Darwin?

There is an unintentional injustice related to both Darwin and plants. When we think of this great scientist and naturalist, we immediately think of Galapagos finches and tortoises, the struggle for existence, and the fundamental question of evolution or creation. However, the general public, as well as university students, may not be as familiar with Darwin's work regarding plants, which is at least as significant as his more famous work, and much more voluminous. Few imagine that in formulating his evolutionary theory,

Darwin's observations and experiments on plants played an equally significant role. This work has been overlooked not only in the multitude of articles published on the occasion of this anniversary, but over time as well. It is another illustration of prejudice, another "blind spot" concerning plants.

Darwin did not regularly attend a structured course leading to a degree that would pave the way for a formal "vocation". He came from a well-to-do family with a tradition in the sciences, both as professionals (his father was a physician) and amateur hobbyists (his grandfather was an excellent naturalist). Darwin's family wanted him to become a physician, so in 1825 he enrolled at the University of Edinburgh at age 16. However, he never received a degree, not because of inconsistency or indolence but perhaps because of his explicitly expressed aversion to the sight of blood and pain, which distanced him from medical content lectures. Yet, nothing stopped him from collecting animals and plants, an activity he had taken up when he was a boy, to the great annoyance of his father. Darwin approached various professors to discuss his observations, which were so detailed and careful that in 1827, at age 18, the indifferent medical student saw his name printed for the first time in a scientific publication. Professor Grant, in his article in the *Edinburgh Journal of Science* on the reproduction of marine invertebrates, acknowledged Darwin thusly: "...with the help of my ardent young friend C. Darwin, who kindly made available to me samples of the egg-cases of *Pontobdella muricata*." Volumes of publications followed in which Darwin's name appeared, and not only in the acknowledgments.

Darwin's failure to become a physician did not daunt his father in seeking another dignified vocation worthy of their social class. The priesthood was a good alternative at that time, but it required a university degree. Apparently, young Darwin accepted the idea, and not only to satisfy his father's wishes. Theologic studies would provide him with the opportunity to remain within an academic environment. Besides, the prevailing view at the time was that science had a duty to apply its methods to prove the wisdom of the Creator. This provided a very good pretext allowing scientists to make observations using their judgment and to perform experiments without social friction – at least to the extent they did not dispute theologic dogmas. It is well known, however, that Darwin did not avoid such societal disapproval, because his evolutionary theory could not leave out *Homo sapiens*. The inevitable hypothesis that humans and primates (gorillas, chimpanzees, etc.) share common ancestors was analysed by Darwin himself in his book *The Origin of Man*, which was published in 1871, 12 years after publication of his monumental work, *The Origin of the Species by Means of Natural Selection*. Naturally, for his era, this hypothesis aroused many protests, but it soon was accepted by the scientific community.

Therefore, to comply with his father's wishes, young Darwin found himself at Cambridge studying theology – in a manner of speaking, of course. He was not a good student, as he never stopped being a devoted and indefatigable naturalist. He applied himself fervently and with exceptional diligence to collecting insects. His name appeared for the second time in a four-volume book on entomology by Professor Stephens: appearing on one page is the reference "captured by C. Darwin." As Darwin wrote later, he felt like a poet who saw his first poem published: two words and his name. Besides collecting, he attended botany classes held by Professor Henslow and geology classes by Professor Sedwick. Darwin also accompanied and helped them with their fieldwork, although this was not part of a student's normal obligations. These two professors appreciated their young associate's discerning intellect, astuteness, enthusiasm, and hard work, and finally accepted their mentoring role. Particularly in the case of Henslow, who was a young man at the time, a long-term cooperation and friendship formed, proving – with the help of some luck – to be a critical point in Darwin's future course and the course of modern biology.

Still, Darwin had to obtain a degree at some point. He rolled up his sleeves and passed his final exams in 1831. He never became a priest. A few months after earning his degree, he received a letter from Henslow informing him that the *HMS Beagle* was to set out on a long exploratory voyage around the world. There was a vacancy on board for a naturalist, for which Henslow would recommend Darwin unconditionally. Darwin's task would be to record his observations on the flora, fauna, and geology of the *terrae incognitae* encountered and to collect biological and geologic material. The young man must have been over the moon; he already was enchanted by the exploratory descriptions of von Humboldt and had dreamt often about such travels. With the help of an equally daydreaming uncle on his mother's side, Darwin managed to break his father's resistance. In December 1831, at only 22 years of age, he embarked on the most famous voyage in the history of science. At that time it was a risky venture for daredevils: the trip had been scheduled for 2 years, lasted 5, and took the group to the Canary Islands, Brazil, the Falkland Islands, Tierra del Fuego and Patagonia in Chile, the Galapagos Islands, the South Seas, Australia, South Africa, the Cape of Good Hope, and other destinations that were exotic even by today's standards. As the ship mapped the coastline, Darwin wandered indefatigably, for months at a time, collecting materials from the hinterlands, making notes, and getting to know places, people, and customs totally different from those back home. While on board, to the extent his annoying seasickness allowed, he packed materials to send to England, mainly to Henslow and other professors. With his samples, he included detailed, clear, and concise notes, and he supported his findings daringly: he formulated hypotheses and proposed interpretations always on the basis of the information already available. Despite the difficulties at that time, particularly while at sea, Darwin read a great deal. Henslow and other recipients of Darwin's dispatches disseminated his discoveries, and Darwin's fame as a significant naturalist spread throughout England even before he and the *Beagle* returned in October 1836.

In the proper hands, this voluminous material had much to reveal. However, it had to be classified, analysed, discussed, enriched with new observations, compared with older findings, recorded, and, finally, published. By that time, Darwin's father had accepted the fact that his son was not cut out to be a physician or priest and left him alone, allowing Darwin to play with his collections and notes and continue his scientific work. The elder Darwin even left Charles a share of the family property. In effect, Darwin's work on evolution was financed by private funds; he never became a university professor. His material was processed not on the laboratory bench of a university but in the rooms and garden of his home. Darwin never received a salary for this work – he lived on family funds. He had no assistants or students to undertake routine chores and tasks. When the going was tough, he recruited his children to help with experiments. One of them, his son Francis, went on to become an important biologist specialising in plant functions. Charles continued to work and publish until his death in 1882. His last book, on the role of earthworms in forming and fertilising soil, was published a few months before his death.

Darwin's biography may appear strange compared with the profile of the average modern researcher, who often is a state worker or, in more scientifically developed countries, an employee of a company or industry. Depending on their fame and effectiveness (often despite their effectiveness), such researchers have at their service assistants and students, who make up their *research team*. They have an office that may be luxurious, with a laboratory containing modern equipment. They usually complain about the shortage of research funds, although none can definitively answer whether a generous increase in grant money would lead to a corresponding advance in their achievements. If a grant is reduced, these scientists usually limit their research activity, as though scientific issues spontaneously disappear. They are preoccupied with being promoted to professor, and when they achieve this landmark, they usually give up research. At some point they retire. Some may claim that Darwin did not have a livelihood problem. Indeed he did not. Not only his own but also his wife's family was well-off. However, few prosperous individuals today would turn their homes into laboratories and their gardens into experimental fields. Others also may claim that research is expensive nowadays. Indeed it is – *mutatis mutandis*. However, in the books in which Darwin describes his experiments, it is worth noting the manner in which he secured devices, instruments, and chemicals. He either borrowed them or accepted donations. When he had an idea how an experiment should be performed and what he actually needed, he visited his university professor friends, who contributed by donating or lending appliances and chemical compounds, when these were necessary. Most often, Darwin needed extremely simple materials, as the most important and decisive role in his experiments was played by an idea rather than by expensive devices. In his writings, one may discern a strong inner drive to conclude an experiment, to learn the truth by devising appliances using everyday materials. Darwin raised questions with a childlike curiosity and applied himself to answering them with an amateur's enthusiasm, which is typical of all good professionals. For a decade, he studied the paradoxic physiology of insectivorous plants. The joy of discovery was such that he did not hesitate to write to his scientist friends that these unusual plants were the most fascinating organisms he had encountered. He wrote the same thing about other organisms he studied at different times. Darwin always declared the current subject of his work as the most wonderful, and his letters to friends reveal he was in continuous rapture and wonder at the miracles of nature.

Although Darwin never became a professor, as early as his *Beagle* years he had become a recognised and select member of the scientific community. He regularly published articles in prestigious scientific journals of his time (signed simply *C. Darwin, at Down House, Kent*, his home address), and his announcements at scientific societies were special events. However, the greatest and most important part of his work is in the form of voluminous books. Darwin generally was held in high esteem and was one of only five Englishmen *sine nobilis* buried at public expense in the 19th century. He was presented with several important scientific awards but did not allow himself to be carried away by fame and recognition. He was modest to the point of bashfulness. There is a rumour that when he decided – with his father's consent – to ask for his cousin Emma's hand in marriage, he lost his courage and left without proposing; all he managed to do was briefly describe his budding theory on the origin of species. He succeeded in his second attempt with quite a bit of urging.

All his biographers agree that Darwin was kind and accessible, yet demanding of himself when it came to publishing his research results. His work on plants includes the following books:

1862—*On the Various Contrivances by Which British and Foreign Orchids Are Fertilized by Insects and the Good Effects of Intercrossing* (365 pages)
1865—*The Movement and Habits of Climbing Plants* (118 pages)

1868—*The Variations of Animals and Plants under Domestication* (two volumes; 806 pages total)
1875—*Insectivorous Plants* (462 pages)
1876—*The Effects of Cross and Self Fertilization in the Vegetable Kingdom* (482 pages)
1877—*The Different Forms of Flowers on Plants of the Same Species* (482 pages)
1880—*The Power of Movement in Plants* (593 pages; in cooperation with his son Francis)

Each of these voluminous books is a full dissertation on its specific topic. Each contains all the information available, at least all that was accessible to Darwin through the methods of his time. New questions are raised that have not been answered, and postulations to be examined are presented. Experiments the author performed in his own home, as always, are described with characteristic clarity and essential detail. Protocols of thorough observation, most of which are quantitative, are organized. Results are analysed and conclusions reached. There is a discussion as to whether the results are compatible with the theory of the origin of species and natural selection. The adaptive significance of form and function is examined in minute detail. The author is not satisfied with findings related to one plant species, nor does he allow himself easy-to-make generalisations. For example, he describes the movements of tendrils and similar organs in more than 100 plants. In some of them, the position of the tendrils in space over time is drawn in such detail that one wonders how this could have been studied at a time when there were no cameras. Given that these movements are exceptionally slow, Darwin must have sat still for hours with his eyes fixed on his subject while noting times and positions. Obviously, each book was the result of years, possibly decades, of labour, if one includes the intellectual preparation for the preliminary study of sources, the processing of questions, and the conception of the *modus operandi* to be implemented. In his study of orchid fertilisation by the intervention of insects, Darwin made observations and conducted experiments on all the orchid species he could cultivate in his garden as well as those he found in nearby fields, a total of some tens of orchids. Darwin's industriousness is amazing, particularly considering his motives. He did not work so intensively to get promoted, curry favour with his boss, or make money. It was his inner desire to get to the essence of things that drove him.

One may wonder whether Darwin's observations and conclusions were finally confirmed. The answer is categorically affirmative for the greatest part of his work. Not only have findings been confirmed, but they are still valid. Darwin's experiments concerning plants' tendency to turn towards light (*phototropism*) are still quoted almost in their entirety in modern textbooks of plant physiology. However, few students are informed about the authorship of the revolutionary – for his time – hypothesis formulated by Darwin. He suggested that the phenomenon cannot be interpreted without recognizing the involvement of hormones. These hormones – like an internal signal – determine the bending, being, in essence, the translation of an environmental light signal that could be meaningful at the cellular level. This hypothesis was confirmed several decades later and was improved recently using modern molecular biology techniques. In addition, very few learn that the paradoxic phenomena used by orchid flowers to deceive insect pollinators were first detected by Darwin, interpreted by him based on flower morphology, and recently complemented by findings on the additional participation of deceitful pheromones — volatile substances released by flowers, the chemical structures and properties of which imitate female insect odours, which are attractive to the male of the insect species involved.

Volumes may be written about Darwin's contribution to modern plant biology, but this is not our goal.

Let us simply accept that we are safely treading on the wealth of his observations; to a great extent we are still tracing Darwin's course. Returning to the original question, what else characterized Darwin as a scientist?

Much has been written about the alleged reasons Darwin delayed publication of his most important book, the one about the origin of species. From his correspondence with the most acclaimed professors of the time, Darwin apparently conceived the basic notion – the triptych "variability [the traits of individuals within the species], natural selection [the most successful individuals leave more descendants] and inheritance [of corresponding traits] and, therefore, increased frequency of these traits in a population, which under suitable conditions generates new species" – long before it was finally published in 1859. Many contend that its publication would have been delayed further had Darwin not received, in 1858, a letter from A. R. Wallace, another important naturalist of the time, who was working in southeast Asia. Scattered papers by Wallace that had already been published indicated he eventually would have reached the same conclusions as Darwin; his 1858 letter leaves no doubt. Two scientists, independently of each other, had reached the same revolutionary conclusions that – although unpublished in their final form – already were known within academic circles. Indisputably, Darwin's conclusions were more comprehensive. Because Darwin was facing family problems at the time – one of his children was sick (and finally died) – and his own health was fragile, his friends from academia (Geology Professor Lyell and Botany Professor Hooker) arranged a joint, albeit brief, presentation at the Linnaean Society of the work of both Darwin and Wallace. It was a fair act so that neither would have to be pushed to "win the race." Although he was so close, Darwin could not make the presentation. Despite his problems, however, he got to work on his book on the origin of species and it was published a year later.

This brief history impressed me for two reasons. When I was a young postgraduate student, I felt pressured by my environment to quickly publish my study results "in case someone else made it first." In time, I became convinced that this rarely happens. So what was Darwin waiting for? Was he lazy? None of his biographers implies this; quite the contrary, Darwin was exceptionally industrious. A short analysis of his work with plants offers an explanation for this apparent delay. It seems Darwin would not risk publication without first "verifying" his arguments. Who today would examine tens of climbing plants, who would examine their structure and function concurrently, who would try to associate his or her findings with the overall behaviour of the plants under study, and, finally, who would wait for a decade before deciding the work was worth publishing? In my view, this is a person who shows genuine interest in scientific truth, who keeps questioning his or her own findings, who is not eager to be promoted or to show off scientifically, who embarks on a journey to enjoy exploration for its own sake not merely to reach a destination. A researcher must not only be free but *feel* free. Darwin appears not to have had any other commitments besides his self-commitment that what he finally put forward was more than likely true. He played without being anxious about time passing by.

The second impressive aspect of this brief history is the publicity Darwin's views enjoyed long before they were published. Darwin announced his findings in letters and discussed them with his professor friends. Wallace exchanged letters with Darwin, and privately they informed each other in detail about their views; ideas circulated freely before they were officially submitted in written form. What a difference from the half-truths and weasel words usually exchanged at present-day scientific conferences and from the secretiveness among scientists at some laboratories, which is broken only after their paper has been published. Some might say the mass research undertaken today heightens the risk of losing the race. It may be argued, though, that this risk is negligible compared with the much more serious hazard of being hasty and disrupted under the effects of stress. If we are interested in the journey rather than the destination, we should let others travel too. Let us enjoy the research process as play, as Darwin did, and let others play as well.

There may be one more point that emerges from studying Darwin's life. The books on plant behaviour presented in the preceding list are the product of the second period of his scientific career, which was focused on the study of functions sealed with experimentation. The first period, which produced *The Origin of the Species* and later *The Origin of Man*, was based on careful observation of nature and organisms. This observation, this close relationship, full of wonder and admiration for the world around him, never left Darwin. Recording things and examining their state were Darwin's life work, which he religiously and devoutly maintained in the second period of his research, characterized by intense study of the physiology of organisms. The topics of his research always stem from his wonder at the real world and its functioning. He never passed any of it by indifferently; he never stopped thinking about the world. In his home in Kent, one may still find the well-known "thinking path" where he used to walk and – I suppose – reorganize his thoughts and observe the world. He observed animals and humans when travelling, even his children at various ages, and recorded their behaviour patterns and expressions, resulting in the publication in 1872 of an extremely efficient and entertaining book, even for those just browsing through it, titled *The Expression of the Emotions in Man and Animals*.

Darwin was scientifically active until the end of his life. His last book, on the importance of earthworms in the formation and fertilization of soil, was published a year before his death. His last paper, "On Instinct," was read at the Linnaean Society a year after his death. It is worth reading, without any comment, his last scientific words:

> *It may not be logical, but to my imagination it is far more satisfactory to look at the young cuckoo ejecting its foster-brothers, ants making slaves, the larvae of Ichneumonidae feeding within the live bodies of their pray, cats playing with mice, otters and cormorants with living fish, not as instincts specially given by the Creator, but as small parts of one general law leading to the advancement of all organic bodies, namely, Multiply, Vary, let the strongest Live and the weakest Die.*

Let us venture a final question: What do you think the reaction of an average fellow citizen would be if, when walking in the countryside, he or she came across a venerable gentleman observing with wonder a beetle crawling around a flower? What would our fellow citizen do if the old gentleman took out his notebook to record the movements of the insect or to note whether the beetle passed through the stamen area? If he used his chronometer to measure how long the beetle remained in the flower? If he smiled occasionally at what he saw? If he picked up the beetle and tried to determine under a magnifying glass whether it was carrying pollen? If he then placed the beetle carefully onto the ground? If the old gentleman often visited the same area, and practiced similarly odd activities? If someone informed our co-citizen that the gentleman in question is not getting paid for this activity but pays for it out of his own pocket and, indeed, enjoys himself? That when the old gentleman eats and sometimes while he is asleep, his mind is occupied by the behaviour of the beetle and its consequences for the plant? And all this is driven by mere curiosity? The response would be the same that drives those – whether young or old – with a similar passion today to hide for fear of becoming victims of derision.

Despite the potential for ridicule, however, there are quite a few people today who practice similar activities with equal passion, dedication, industriousness, curiosity, consistency, a keen eye for observation, enthusiasm, and imagination, regardless of low pay, because their precious reward is the joy of play and discovery. To scientists with these characteristics, we owe the knowledge we already have of the world and the possibility of understanding it better in the future. To scientists with these characteristics, of a lower or higher calibre, we owe our views about life and plants today, however incomplete these may be. I am certain, though, that Darwin, with his specific character traits, might have become an equally competent physician, priest, engineer, carpenter, farmer, or cobbler, and would have led an equally fulfilling life. This leads to the conclusion that a potential scientist, that is, one who is "*-scient*," who knows in depth the object of his vocation, may be found anywhere around us: in the office, the underground, or a hospital or factory; behind a laboratory or a cottage-industry bench; in an amphitheatre; or driving a tractor. You will recognize them instantly because, like children, they never tire of asking questions about the nature of even the simplest things, because when they formulate a question, their eyes are full of wonder. When they receive or discover a reasonable answer, wonder turns to admiration and spontaneous joy. Such a person might well be you, reading these words.

Additional Reading

Books

Bazzaz FA (1996) Plants in changing environments: linking physiological, population and community ecology. Cambridge University Press, New York

Beerling D (2007) The emerald planet: how plants changed Earth's history. Oxford University Press, Oxford

Breckle SW (2002) Walter's vegetation of the Earth: the ecological systems of the geo-biosphere. Springer, Berlin

Dawson J, Lucas R (2005) The nature of plants: habitats, challenges and adaptations. CSIRO Publishing, Collingwood

Harborne J (1993) Introduction to ecological biochemistry. Academic Press, London

Harper JL (1977) Population biology of plants. Academic Press, London

King J (1997) Reaching for the sun: how plants work. Cambridge University Press, Cambridge

Larcher W (2003) Physiological plant ecology: ecophysiology and stress physiology of functional groups. Springer, Berlin

Levetin E, McMahon K (2005) Plants and society. McGraw-Hill, New York

Margulis L, Schwartz KV (1997) Five kingdoms: illustrated guide to the phyla of life on earth. WH Freeman, New York

Mauseth JD (2009) Botany: an introduction to plant biology. Jones and Bartlett Publishers, Boston

Niklas KJ (1997) The evolutionary biology of plants. The University of Chicago Press, Chicago

Proctor M, Yeo P, Lack A (2009) The natural history of pollination. Harper Collins Publishers, London

Ridge I (2002) Plants. Oxford University Press, New York

Rothschild LJ, Lister AM (2003) Evolution on planet earth: the impact of the physical environment. Elsevier, Burlington

Schoonhoven LM, van Loon JJA, Dicke M (2006) Insect-plant biology: from physiology to evolution. Oxford University Press, Oxford

Taiz L, Zeiger E (2006) Plant physiology. Sinauer Associates, Sunderland

Willis KJ, McElwain JC (2002) The evolution of plants. Oxford University Press, New york

Reviews, Opinions, and Research Papers

Alvarez LW, Alvarez W, Asaro F et al (1980) Extraterrestrial cause for the Cretaceous-Tertiary extinction – Experimental results and theoretical interpretation. Science 208:1095–1108

Anstett MC, Hossaert-McKey M, Kjellberg F (1997) Figs and fig pollinators: evolutionary conflicts in a coevolved mutualism. Trends Ecol Evol 12:94–99

Austin JJ, Smith AB, Thomas RH (1997) Palaeontology in a molecular world: the search for authentic ancient DNA. Trends Ecol Evol 12:303–306

Bakker RT (1978) Dinosaur feeding-behavior and origin of flowering plants. Nature 274:661–663

Baldwin IT, Halitschke R, Paschold A et al (2006) Volatile signaling in plant-plant interactions: "Talking trees" in the genomics era. Science 311:812–815

Ballaré CL (1999) Keeping up with the neighbours: phytochrome sensing and other signalling mechanisms. Trends Plant Sci 4:97–102

Ballaré CL (2001) Smart plants or stealthy bugs? Trends Plant Sci 6:142

Ballaré CL, Scopel AL, Jordan J et al (1994) Signaling among neighboring plants and the development of size inequalities in plant populations. Proc Natl Acad Sci USA 91:10094–10098

Barthlott W, Porembski S, Fischer E et al (1998) First protozoa-trapping plant found. Nature 392:447

Barrett PM, Willis KJ (2001) Did dinosaurs invent flowers? Dinosaur-angiosperm coevolution revisited. Biol Rev 76:411–447

Bateman RM, Crane PR, DiMichele WA et al (1998) Early evolution of land plants: phylogeny, physiology, and ecology of the primary terrestrial radiation. Annu Rev Ecol Syst 29:263–292

Bendall DS, Howe CJ, Nisbet EG et al (2008) Photosynthetic and atmospheric evolution. Philos Trans R Soc Lond B Biol Sci 363:2625–2628

Bennett KD, Tzedakis PC, Willis KJ (1991) Quaternary refugia of North European trees. J Biogeogr 18:103–115

Benton MJ, Twitchett RJ (2003) How to kill (almost) all life: the end-Permian extinction event. Trends Ecol Evol 18:358–365

Berner RA (1997) The rise of plants and their effect on weathering and atmospheric CO2. Science 276:544–546

Bidartondo MI, Burghardt B, Gebauer G et al (2004) Changing partners in the dark: isotopic and molecular evidence of ectomycorrhizal liaisons between forest orchids and trees. Philos Trans R Soc Lond B Biol Sci 271:1799–1806

Bidartondo MI, Redecker D, Hijri I et al (2002) Epiparasitic plants specialized on arbuscular mycorrhizal fungi. Nature 419:389–392

Bond WJ, Keeley JE (2005) Fire as a global 'herbivore': the ecology and evolution of flammable ecosystems. Trends Ecol Evol 20:387–394

Bond WJ, Midgley GF, Woodward FI (2003) The importance of low atmospheric CO2 and fire in promoting the spread of grasslands and savannas. Glob Chang Biol 9:973–982

Bowman D (2005) Understanding a flammable planet – climate, fire and global vegetation patterns. New Phytol 165:341–345

Briffa KR, Shishov VV, Melvin T et al (2008) Trends in recent temperature and radial tree growth spanning 2000 years across northwest Eurasia. Philos Trans R Soc Lond B Biol Sci 363:2271–2284
Brocks JJ, Logan GA, Buick R et al (1999) Archean molecular fossils and the early rise of eukaryotes. Science 285:1033–1036
Brown TA (1999) How ancient DNA may help in understanding the origin and spread of agriculture. Philos Trans R Soc Lond B Biol Sci 354:89–97
Caldeira K, Kasting JF (1992) The life span of the biosphere revisited. Nature 360:721–723
Callaway RM, Pennings SC, Richards CL (2003) Phenotypic plasticity and interactions among plants. Ecology 84:1115–1128
Chaloner WG, McElwain J (1997) The fossil plant record and global climatic change. Rev Palaeobot Palynol 95:73–82
Chapman RF (2003) Contact chemoreception in feeding by phytophagous insects. Annu Rev Entomol 48:455–484
Chyba CF, Thomas PJ, Brookshaw L et al (1990) Cometary delivery of organic molecules to the early Earth. Science 249:366–373
Cockell CS, Raven JA (2007) Ozone and life on the Archaean Earth. Philos Trans R Soc A Math Phys Eng Sci 365:1889–1901
Čufar K (2007) Dendrochronology and past human activity – a review of advances since 2000. Tree-Ring Res 63:47–60
Després L, David JP, Gallet C (2007) The evolutionary ecology of insect resistance to plant chemicals. Trends Ecol Evol 22:298–307
Douglas SE (1998) Plastid evolution: origins, diversity, trends. Curr Opin Genet Dev 8:655–661
Edwards W, Moles AT (2009) Re-contemplate an entangled bank: the power of movement in plants revisited. Bot J Linn Soc 160:111–118
Ellison AM, Gotelli NJ (2009) Energetics and the evolution of carnivorous plants – Darwin's 'most wonderful plants in the world'. J Exp Bot 60:19–42
Falik O, Reides P, Gersani M et al (2003) Self/non-self discrimination in roots. J Ecol 91:525–531
Falkowski PG, Godfrey LV (2008) Electrons, life and the evolution of Earth's oxygen cycle. Philos Trans R Soc Lond B Biol Sci 363:2705–2716
Falkowski PG, Isozaki Y (2008) The story of O_2. Science 322:540–542
Franco M (1986) The influence of neighbors on the growth of modular organisms with an example from trees. Philos Trans R Soc Lond B Biol Sci 313:209–225
Gersani M, Abramsky Z, Falik O (1998) Density-dependent habitat selection in plants. Evol Ecol 12:223–234
Gray J, Shear W (1992) Early life on land. Am Sci 80:444–456
Grime JP, Mackey JML (2002) The role of plasticity in resource capture by plants. Evol Ecol 16:299–307
He XH, Critchley C, Bledsoe C (2003) Nitrogen transfer within and between plants through common mycorrhizal networks (CMNs). Crit Rev Plant Sci 22:531–567
Herre EA, Jandér KC, Machado CA (2008) Evolutionary ecology of figs and their associates: recent progress and outstanding puzzles. Annu Rev Ecol Evol Syst 39:439–458

Hetherington AM, Woodward FI (2003) The role of stomata in sensing and driving environmental change. Nature 424:901–908
Holopainen JK (2004) Multiple functions of inducible plant volatiles. Trends Plant Sci 9:529–533
Howe CJ, Barbrook AC, Nisbet RER et al (2008) The origin of plastids. Philos Trans R Soc Lond B Biol Sci 363:2675–2685
Huber-Sannwald E, Pyke DA, Caldwell MM (1997) Perception of neighbouring plants by rhizomes and roots: morphological manifestations of a clonal plant. Can J Bot 75:2146–2157
Igamberdiev AU, Lea PJ (2006) Land plants equilibrate O2 and CO2 concentrations in the atmosphere. Photosynth Res 87:177–194
Izaguirre MM, Mazza CA, Biondini M et al (2006) Remote sensing of future competitors: impacts on plants defenses. Proc Natl Acad Sci USA 103:7170–7174
Jacobs BF, Kingston JD, Jacobs LL (1999) The origin of grass-dominated ecosystems. Ann Mo Bot Gard 86:590–643
Kelly CK (1990) Plant foraging: a marginal value model and coiling response in *Cuscuta subinclusa*. Ecology 71:1916–1925
Kleidon A, Fraedrich K, Heimann M (2000) A green planet versus a desert world: Estimating the maximum effect of vegetation on the land surface climate. Clim Chang 44:471–493
Knoll AH (1996) Breathing room for early animals. Nature 382:111–112
Koch GW, Stillet SC, Jennings GM et al (2004) The limits to tree height. Nature 428:851–854
Kouki M, Manetas Y (2002) Toughness is less important than chemical composition of *Arbutus* leaves in food selection by *Poecilimon* species. New Phytol 154:399–407
Kouwenberg LLR, McElwain JC, Kurschner WM et al (2003) Stomatal frequency adjustment of four conifer species to historical changes in atmospheric CO2. Am J Bot 90:610–619
Kutschera U, Niklas KJ (2004) The modern theory of biological evolution: an expanded synthesis. Naturwissenschaften 91:255–276
Labandeira CC, Dilcher DL, Davis DR et al (1994) 97-Million years of angiosperm-insect association-paleobiological insights into the meaning of coevolution. Proc Natl Acad Sci USA 91:12278–12282
Levizou E, Karageorgou P, Psaras GK et al (2002) Inhibitory effects of water soluble leaf leachates from *Dittrichia viscosa* on lettuce root growth, statocyte development and graviperception. Flora 197:152–157
Li P, Bohnert HJ, Grene R (2007) All about FACE – plants in a high [CO2] world. Trends Plant Sci 12:87–89
Liste HH, White JC (2008) Plant hydraulic lift of soil water – implications for crop production and land restoration. Plant and Soil 313:1–17
Macías FA, Oliveros-Bastidas A, Marín D et al (2008) Plant biocommunicators: their phytotoxicity, degradation studies and potential use as herbicide models. Phytochem Rev 7:179–194
Mallik MAB, Williams RD (2005) Allelopathic growth stimulation of plants and microorganisms. Allelopathy J 16:175–198
Manetas Y (2003) The importance of being hairy: the adverse effects of hair removal on stem photosynthesis of *Verbascum speciosum* are due to solar UV-B radiation. New Phytol 158:503–508
Marchand PJ (2000) Seeds of fortune. Natur Hist 109:28–29

Margulis L (1990) Words as battle cries – symbiogenesis and the new field of endocytobiology. Bioscience 40:673–677
Margulis L, Sagan D (2001) The beast with five genomes. Natur Hist 110:38–41
Marx J (2004) Remembrance of winter past. Science 303:1607
McConnaughay KDM, Bazzaz FA (1991) Is physical space a soil resource? Ecology 72:94–103
McElwain JC, Punyasena SW (2007) Mass extinction events and the plant fossil record. Trends Ecol Evol 22:548–557
Meyer-Berthaud B, Decombeix AL (2007) A tree without leaves. Nature 446:861–862
Molas ML, Kiss JZ (2009) Phototropism and gravitropism in plants. Adv Bot Res 49:1–34
Moran J (2006) Life and death in a pitcher. Natur Hist 115:56–62
Morison JIL (1987) Plant growth and CO2 history. Nature 327:560
Muller J (1984) Significance of fossil pollen for angiosperm history. Ann Mo Bot Gard 71:419–443
Musser RO, Hum-Musser SM, Eichenseer H et al (2002) Caterpillar saliva beats plant defences – a new weapon emerges in the evolutionary arms race between plants and herbivores. Nature 416:599–600
Niklas KJ, Spatz HC, Vincent J (2006) Plant biomechanics: an overview and prospectus. Am J Bot 93:1369–1378
Nisbet EG, Cann JR, Vandover CL (1995) Origins of photosynthesis. Nature 373:479–480
Nisbet EG, Fowler CMR (1996) Early life – some liked it hot. Nature 382:404–405
Nisbet EG, Nisbet RER (2008) Methane, oxygen, photosynthesis, rubisco and the regulation of the air through time. Philos Trans R Soc Lond B Biol Sci 363:2745–2754
Nisbet EG, Sleep NH (2001) The habitat and nature of early life. Nature 409:1083–1091
Novoplansky A, Cohen D, Sachs T (1990) How portulaca seedlings avoid their neighbors. Oecologia 82:490–493
Olson JM (2006) Photosynthesis in the Archean era. Photosynth Res 88:109–117
Pagani M, Freeman KH, Arthur MA (1999) Late Miocene atmospheric CO2 concentrations and the expansion of C-4 grasses. Science 285:876–879
Peak D, West JD, Messinger SM et al (2004) Evidence for complex, collective dynamics and emergent, distributed computation in plants. Proc Natl Acad Sci USA 101:918–922
Peñuelas J (2005) A big issue for trees. Nature 437:965–966
Primack R (1999) Life in bloom. Natur Hist 108:56–59
Purdie R (2009) Charles Darwin's plant biology. Funct Plant Biol 36:481–489
Raguso RA (2008) Wake up and smell the roses: the ecology and evolution of floral scent. Annu Rev Ecol Evol Syst 39:549–569
Ratzka A, Vogel H, Kliebenstein DJ et al (2002) Disarming the mustard oil bomb. Proc Natl Acad Sci USA 99:11223–11228
Raven JA (1993) The evolution of vascular plants in relation to quantitative functioning of dead water-conducting cells and stomata. Biol Rev 68:337–363

Raven JA (2008) Transpiration: how many functions? New Phytol 179:905–907
Raven JA, Edwards D (2001) Roots: evolutionary origins and biogeochemical significance. J Exp Bot 52:381–401
Roberts L (1989) How fast can trees migrate? Science 243:735–737
Rothschild LJ (2008) The evolution of photosynthesis. . .again? Philos Trans R Soc Lond B Biol Sci 363:2787–2801
Rumpho ME, Summer EJ, Manhart JR (2000) Solar-powered sea slugs. Mollusc/algal chloroplast symbiosis. Plant Physiol 123:29–38
Schenk HJ, Callaway RM, Mahall BE (1999) Spatial root segregation: are plants territorial? Adv Ecol Res 28:145–180
Schulze-Makuch D, Irwin LN (2006) The prospect of alien life in exotic forms on other worlds. Naturwissenschaften 93:155–172
Selosse MA, Richard F, He X et al (2006) Mycorrhizal networks: des liaisons dangereuses? Trends Ecol Evol 21:621–628
Shenmiller J, Mudgett MB, Schopf JW et al (1995) Exceptional seed longevity and robust growth – ancient sacred Lotus from China. Am J Bot 82:1367–1380
Shepherd RW, Wagner GJ (2007) Phylloplane proteins: emerging defenses at the aerial frontline? Trends Plant Sci 12:51–56
Shulaev V, Silverman P, Raskin I (1997) Airborne signalling by methyl salicylate in plant pathogen resistance. Nature 385:718–721
Shurin JB (2006) A pitcher of things to come. Nature 443:399–400
Silvertown J, Gordon DM (1989) A framework for plant behaviour. Annu Rev Ecol Syst 20:349–366
Simard SW, Perry DA, Jones MD et al (1997) Net transfer of carbon between ectomycorrhizal tree species in the field. Nature 388:579–582
Stebbins GL (1981a) Coevolution of grasses and herbivores. Ann Mo Bot Gard 68:75–86
Stebbins GL (1981b) Why are there so many species of flowering plants. Bioscience 31:573–577
Sterck L, Rombauts S, Vandepoele K et al (2007) How many genes are there in plants (. . .and why are they there)? Curr Opin Plant Biol 10:199–203
Sultan SE (2000) Phenotypic plasticity for plant development, function and life history. Trends Plant Sci 5:537–542
Summers A (2005) How trees get high. Natur Hist 114:34–35
Szathmáry E, Jordán F, Pál C (2001) Can genes explain biological complexity? Science 292:1315–1316
Taylor JE, McAinsh MR (2004) Signalling crosstalk in plants: Emerging issues. J Exp Bot 55:147–149
Tolbert NE (1997) The C2 oxidative photosynthetic carbon cycle. Annu Rev Plant Biol 48:1–25
Traverse A (1988) Plant evolution dances to a different beat: plant and animal evolutionary mechanisms compared. Hist Biol 1:277–301
Traverse A (1990) Plant evolution in relation to world crises and the apparent resilience of Kingdom Plantae. Glob Planet Change 82:203–211
Trewavas A (2005) Plant intelligence. Naturwissenschaften 92:401–413
Trewavas A (2009) What is plant behaviour? Plant Cell Environ 32:606–616

Van Der Heijden MGA, Klironomos JN, Ursic M et al (1998) Mycorrhizal fungal diversity determines plant biodiversity, ecosystem variability and productivity. Nature 396:69–72
Vandermeer J, Odling-Smee FJ, Laland KN et al (2004) Niche construction – the neglected process in evolution. Science 303:472–474
Wandersee JH, Schussler EE (1999) Preventing plant blindness. Am Biol Teach 61:84–86
Willis KJ, Bennett KD (1995) Mass extinction, punctuated equilibrium and the fossil plant record. Trends Ecol Evol 10:308–309
Woese CR, Kandler O, Wheelis ML (1990) Towards a natural system of organisms: proposal for the domains Archaea, Bacteria and Eucarya. Proc Natl Acad Sci USA 87:4576–4579
Wolfe DW (2001) Out of thin air. Natur Hist 110:44–53

Index

CPSIA information can be obtained at www.ICGtesting.com
Printed in the USA
LVOW101531290313

326723LV00008B/149/P

9 783642 283376